ON THE PRACTICE OF SAFETY

ON THE PRACTICE OF SAFETY

Second Edition

Fred A. Manuele, CSP, PE

JOHN WILEY & SONS, INC.

New York • Chichester • Weinheim • Brisbane • Singapore • Toronto

Originally published as ISBN 0-442-02423-1

Published simultaneously in Canada.

For ordering and customer service, call 1-800-CALL-WILEY.

Library of Congress Cataloging-in-Publication Data

Manuele, Fred A.
On the practice of safety / Fred A. Manuele.--2nd ed.
p. cm.
Includes bibliographical references and index.
ISBN 0-471-29213-3
I. Title.
T55.M353 1997
658.4'08--dc21 97-26554
CIP

Printed in the United States of America.

10 9 8 7 6 5 4 3 2

To Irene

Contents

Preface to the Second Edition *ix*
Preface to the First Edition *xi*
Acknowledgments *xiii*
Introduction *xv*

1. Transitions Affecting the Practice of Safety 1
2. Principles for the Practice of Safety: A Basis for Discussion 15
3. Defining the Practice of Safety 27
4. Academic and Skill Requirements for the Practice of Safety 35
5. On Becoming a Profession 45
6. Observations on Causation Models for Hazards-Related Incidents 56
7. A Systemic Causation Model for Hazards-Related Occupational Incidents 74
8. Incident Investigation: Studies of Quality 83
9. Designer Incident Investigation 93
10. Applied Ergonomics: Significance and Opportunity 114
11. System Safety: The Concept 134
12. Safety Professionals and the Design Processes 145
13. Guidelines: Designing for Safety 157
14. Comments on Hazards and Risks 175
15. Hazard Analysis and Risk Assessment 188
16. On Quality Management and the Practice of Safety 202
17. On Safety, Health, and Environmental Audits 218
18. Anticipating OSHA's General Industry Safety and Health Program Standard and a Safety Management Audit System 225

19. Successful Safety Management: A Reflection of an Organization's Culture 235
20. Measurement of Safety Performance 251

Index 267

Preface to the Second Edition

Four-plus years have passed since the first edition of this book was published. Transitions in the practice of safety have continued to take place, requiring some revisions in all of the essays originally written and the addition of new ones. These are the new or significantly revised chapters.

"Transitions Affecting the Practice of Safety" is the first chapter in this edition and it replaces the first chapter in the previous edition. It addresses the changes that have taken place in the business climate and their effect on career planning. It also discusses how those changes require having a practice that has a sound theoretical and practical base, and the opportunities they present. This new chapter also includes the substance of the last chapter in the first edition, which was "On Management Fads."

"Principles for the Practice of Safety: A Basis for Discussion," Chapter 2, should be of interest to those who think about the substance of what safety professionals do and want to move the state of the art forward.

In recognition of the need to further our knowledge of hazards-related incident causation, Chapter 7 covers "A Systemic Causation Model for Hazards-Related Occupational Incidents."

"Designer Incident Investigation" is intended to help safety personnel craft practical and realistic incident investigation systems, after considering the resources available and the organizational sophistication.

"Guidelines: Designing for Safety" is a concept paper to assist those whose responsibilities are extended into the design processes.

In the first edition, there were separate chapters offering comments on hazards and risks. They have been combined into one chapter—"Comments on Hazards and Risks."

The first edition chapter "Anticipating OSHA's Standard for Safety and Health Program Management" has been expanded to include comments on the safety management audit system OSHA is exploring.

Having sound safety performance measurement systems has become more important, thus the new chapter "Measurement of Safety Performance."

This second edition emphasizes the need to balance the design and engineering aspects and the management and task performance aspects of the practice of safety.

Preface to the First Edition

Throughout a rewarding career as a safety professional, I have observed with pride the achievements of the many who have contributed to the recognition of the practice of safety as a profession. We've come a long way. It is my hope that through this book I can assist in furthering that progress.

Ours is a profession in transition, with many additional opportunities for professional involvement, accomplishment, and recognition. In light of the transitions in progress, the essays in this book propose an extension of knowledge by those engaged in the practice of safety, through which they can enhance their careers, become more effective, and be perceived as greater contributors toward achieving organizational goals.

This is also a book about fundamental principles and practices. It is addressed to students, educators, and to practicing safety professionals. I hope that readers will welcome a review of our fundamentals, intended to improve the theoretical and practical base for our practice. I have attempted to argue systematically and consistently for hazards management ideas of substance in relation to our history, opportunities, and needs.

I expect that these essays will be slightly controversial since some long-accepted concepts are put to question. Some practices that have become fashionable are not sound. I understand that what I propose will be perceived by some to be heresy. Some readers will presume that I have put their finest ideals under scrutiny and proposed significantly different concepts.

In an atmosphere of quiet deliberation, my best hope is that what I have written will be perceived as thought provoking and will serve to further the professional practice of safety.

Acknowledgments

To properly recognize all who have contributed to this book, I would have to cover a period of seventy-two years and the list of people deserving appreciation would be endless. Particularly, though, to the people who gave of their time in past years and critiqued individual essays, I express my sincere thanks and gratitude.

Introduction

The essays in this book cover a broad range of subjects pertaining to the practice of safety, and I thought it would be helpful to readers—students, educators, and safety professionals—if I provided guidance on their content through a brief synopsis of each of them.

1. Transitions Affecting the Practice of Safety

Performance expectations resulting from changes that have taken place in the business climate in recent years are higher, and somewhat different from before. How safety professionals can respond effectively to those expectations is discussed here. These changes also promote the development of personal career enhancement plans, for which an outline is provided. Opportunities presented by the emerging safety technology are also addressed.

2. Principles for the Practice of Safety: A Basis for Discussion

For the practice of safety to be recognized as a profession, it must have a sound theoretical and practical base. I propose that there is a generic base for the work of safety professionals that must be understood and applied if we are to be effective. We take a variety of approaches to achieve safety, and they can't all be right. To promote discussion, a listing of general principles, statements, and definitions is given.

3. Defining the Practice of Safety

We who call ourselves safety professionals will never be accepted as a profession until we agree on a definition of the practice of the safety, make it known, and meet its requirements. This essay identifies the societal need fulfilled by safety professionals, establishes why safety professionals exist, and sets forth the fundamentals of the practice of safety.

4. Academic and Skill Requirements for the Practice of Safety

Reviews are given of the academic knowledge and skill that would prepare one to enter the practice of safety, and of the knowledge and skill requirements for the applied practice of safety, using the term broadly.

5. On Becoming a Profession

Safety practitioners will attain professional recognition when the practice of safety meets the regimens of a profession. Recognition of the great accomplishments thus far toward achieving professional recognition and a review of our present status, needs, and actions that we should consider are encompassed in a discussion outline.

6. Observations on Causation Models for Hazards-Related Incidents

If several safety professionals investigate a given hazards-related incident, they should identify the same causal factors, with minimum variation. That is unlikely if the thought processes they use have greatly different foundations. At least twenty-five causation models have been published. Since many of them conflict, all of them cannot be valid. A review of some of them is followed by a discussion of principles that should be contained in a causation model.

7. A Systemic Causation Model for Hazards-Related-Incidents

There is a plea in the causation model presented for recognition of the impact an organization's culture has on causal factor development, and for a balanced, systemic approach that addresses both the design and engineering, and the management, operational, and task performance causal factors.

8. Incident Investigation: Studies of Quality

Hazard analysis is the most important safety process. If that process fails, all other processes are likely to be ineffective. Incident investigation serves as one vital basis for hazard analysis. Initially, a collection was made from eleven companies of 537 actual incident investigation reports for a study of their quality. This treatise gives the findings of the study, provides a self-evaluation outline through which the quality of incident investigation can be assessed, and discusses how incident investigation can be improved. A subsequent study was made of the quality of incident investigation in six companies that have superior achievements in safety, and the findings of that study are recorded.

9. Designer Incident Investigation

Because of the variances in safety cultures and the resources available, it is folly to suggest that an incident investigation system could be crafted that would universally apply in all organizations. Guidance is given for a safety professional to assess that which is attainable and to draft an incident investigation system that realistically relates to organizational culture and sophistication.

10. Applied Ergonomics: Significance and Opportunity

Ergonomics has emerged to become a major element in the practice of safety. Successes from many ergonomics revisions in workplace and work methods redesign that were initiated to resolve injury and illness problems have also achieved increases in productivity and reductions in costs. That gets management attention, and provides opportunities for extended involvement by safety professionals and recognition. This essay looks to the future and explores the long-term impact of ergonomics on the content of safety practice—it will be significant.

11. System Safety: The Concept

Lessons can be learned from the successes achieved by system safety practitioners. System safety is hazards- and design-based; so is the entirety of the practice of safety. This essay establishes why it is important for generalist safety professionals to acquire knowledge of system safety principles, and outlines the system safety idea. As opportunities arise for generalist safety professionals to participate in the design processes, the need for system safety skills will be apparent.

12. Safety Professionals and the Design Processes

I believe that the greatest strides forward respecting safety, health, and the environment will be made in the design processes. An awareness of the soundness of that premise is emerging. This is where the first safety decisions are made. Discussions establish why safety professionals should become involved in design processes, how to get there, and the skills needed.

13. Guidelines: Designing for Safety

Since transitions in the practice of safety are leading safety professionals into greater involvement in the design processes, readily available guidelines on the concepts of designing for safety would serve them well. Since a search I made for such guidelines was unproductive, an informational, educational reference paper was written for those who participate in having safety designed into all processes, the workplace, and the work methods—applying proactive and cost-effective concepts.

14. Comments on Hazards and Risks

Hazards are the justification for the existence of all safety professionals—whatever titles they may use. A hazard is defined as the potential for harm: Hazards include the characteristics of things and the actions and inactions of people. This chapter stresses that safety professionals must have an understanding of the duality of the nature of hazards for all aspects of their work, and of the relation of hazards to risks. It is clearly established that all risks with which safety professionals deal derive from hazards. There are no exceptions. Risk, in the context of the work of safety professionals, requires a measure of the probability of hazards-related incidents occurring and of the severity of harm or damage that could result. Thus, a case is made that the two distinct aspects of risk must be considered by safety professionals in fulfilling their responsibilities: avoiding, eliminating, or reducing the *probability* of a hazard being realized; and minimizing the *severity* of adverse results if an incident occurs.

15. Hazard Analysis and Risk Assessment

Safety professionals cannot properly give advice on hazards and risks unless the hazards are analyzed and the risks deriving from them are assessed as to their significance. This essay works through the process of hazard analysis and subsequent risk assessment. It explores hazard analysis and risk assessment methods, discusses the jeopardy in using some published techniques, and concludes with a practical methodology outline.

16. On Quality Management and the Practice of Safety

This chapter establishes the similarities between the principles of quality management and the principles for the practice of safety, explores the theoretical ideal for quality and safety, reviews the criteria for The Malcolm Baldrige National Quality Award, and comments on the work of W. Edwards Deming, Philip B. Crosby, Joseph M. Juran, Frank M. Gryna, Gary L. Winn, and Graham Mark Brown, Darcy E. Hitchcock, and Marsha L. Willard. Comments are also made about the Six Sigma quality management program at Motorola. Why some quality management and safety management initiatives succeed or fail is discussed.

17. On Safety, Health, and Environmental Audits

I believe that most safety, health, and environmental audits, which are to measure the quality of hazards management in place, are deficient in purpose and content. Shortcomings addressed in this chapter are the assessments given to organizational culture, management commitment, design and engineering practices, the inadequate attention given to low probability–high consequence events, and prioritizing risk reduction measures. Principal purposes of an audit, what an audit should accomplish, and the criteria that a safety audit should meet are discussed.

18. Anticipating OSHA's General Industry Safety and Health Program Standard, and a Safety Management Audit System

OSHA has in progress activity to establish a safety and health program standard for general industry, and a formalized audit system to be used by its compliance officers to assess the quality of employer safety and health programs. If such a standard is promulgated and such a safety audit system is adopted, the impact on workplace safety could be greater than that of all of the specific standards OSHA has thus far issued. A review is given of the history leading to what is being proposed, and the model on which the proposed standard and the audit system are based.

19. Successful Safety Management: A Reflection of an Organization's Culture

An organization's culture determines the probability of success of its hazards management endeavors. What the board of directors or senior management decides is acceptable for the prevention and control of hazards is a reflection of its culture. That theme is established in this chapter. Also, results of interviews conducted with safety personnel in organizations where the culture requires highly successful safety performance are set forth, along with comments on how safety management programs are made to work successfully.

20. Measurement of Safety Performance

This essay responds to a renewed interest in having measurement systems that are universally applicable, effectively assess occupational safety performance, and communicate well in terms that managements understand. Several safety measurement systems are discussed.

Chapter 1

Transitions Affecting the Practice of Safety

INTRODUCTION

This essay addresses the changes in the business climate in the past few years, comments on possible future changes, discusses evolving technology, and explores how those transitions affect the practice of safety.

To say that there has been volatility in the employment practices of larger companies in the past few years is an understatement. Downsizing, right-sizing, and dumb-sizing are now commonly used terms. Predictions of the future by business writers vary greatly. A few are predicting less turmoil and more stability. Yet Peter F. Drucker, in *Managing in a Time of Great Change* (1), a November 1995 publication, makes this comment:

> In fact, if you make a wager on any big company, the chances of it being split within the next ten years are better than the chances of it remaining the way it is.

But some professors in universities giving safety degrees say that they do not observe as pervasive a mood of anxiety about employment and careers in their graduates who are employed in small- and medium-size companies as that observed in larger companies by consultant-writers.

Nevertheless, it can be expected with a great deal of certainty that there will be a continuum of economic ups and downs and a never-ending proliferation of management fads—all having an impact on business practices and career potentials. More than in the past, astute professionals will recognize that their interests are best served by being attentive to their marketability and preparing for the reality of the world that is. This requires a continuing introspection concerning their knowledge and skills and keeping current with the evolving technology in the practice of safety. Significantly, they must assess the perceptions others have of the substance and worth of the consultancies they provide.

While the business climate has been volatile, I suggest a positive approach to career potential by safety professionals since a broad range of opportunities exists for expansion of their capabilities, greater effectiveness, and recognition.

Unfortunately, we safety professionals do not have a history of anticipating developing needs and taking the leadership in providing solutions to fulfill those needs. Typically, we are reactors. In the current and foreseeable business climate, we must anticipate what is needed for career security and be initiators of imaginative solutions in relation to evolving technological developments.

Major Influences on the Practice of Safety

In the past few years, the practice of safety has been greatly influenced by four major developments. They are:

- a more demanding business climate that requires greater management effectiveness, a part of which is a stringent measurement of the contributions of individuals toward achieving entity goals;
- the extended impact of applied ergonomics on the practice of safety;
- the realization that quality management principles and safety management principles have a remarkable kinship;
- the frequent bringing together of the functions of safety, health, and environmental affairs under a single management.

Impact of the Transitions in Business Practices

Several authors have addressed the popular management fads, their effectiveness or ineffectiveness, and their impact on employees. These are examples, paraphrased, of what has been written recently.

- Contrary to conventional wisdom, corporations that have used the latest management wrinkles and trends have not become more efficient, flexible, or competitive.
- Empowerment, the buzzword for participatory management that reverberates throughout the boardrooms of corporate America, is not working.
- Restructuring not only fails to create new forms of participation, it destroys old forms of cooperation that kept good bureaucracies working reasonably well in the past.
- Because of the way restructuring and reengineering have been done, the American workforce is filled with legitimate fear and cynicism.

As I did in the first edition of this book, I suggest that safety professionals be alert to the new management fads that surely will be adopted, and produce their own turmoil. Throughout that volatility, their best interests are served by having a sound knowledge and skill base, both in management practices and in safety technology.

Of the several thousand management books published each year, I chose two from which to excerpt. Being opposites in a way, the two are indicative of the great variations that can be expected in management books. One is a real downer; the other is speculatively encouraging.

White-Collar Blues (2) by Charles Heckscher resulted from a study of fourteen large organizations to determine the effects on middle managers of corporate restructuring. These companies had gone through reengineering, reorganizing to achieving flatter organizations, downsizing, right-sizing, or dumb-sizing—as have many safety professionals.

Briefly, Heckscher found that: loyalty, downward or upward, had diminished greatly; pressure is great to work harder, for higher productivity, and for beating out fellow workers; coaching and focus groups are in, but they do not always solve problems; worker participation (empowerment) has become a widely accepted principle but it is less widely implemented in practice; and restructuring and downsizing have actually reinforced corporate bureaucracy instead of eliminating it.

This general observation by Heckscher gives an indication of what the future business climate may be like:

> The problem is that today's business environment, if not the world in general, is characterized by [this] condition: we don't know what we're doing; at best we'll figure it out when we get there. No large company that I know of, and certainly none in my sample, had confidence in even a rough picture of what markets and economic conditions would be in five years. They are searching for a way to be continuously adaptable.

Heckscher observed that "without a shared ethic, groups fragment and individuals feel lost" and that "most people appear to have a powerful need to feel that they are part of and contributing to something larger than themselves." Unfortunately, recent trends indicate that as the loyalty factor diminishes, so also does the shared ethic. Individuals are less committed to any company but, rather, to a personal set of skills, goals, interests, and affiliations.

Heckscher also noted that building a personality that is tough enough and flexible enough to avoid dependence on an organization is a difficult process.

Be assured that the management practices in place today will be replaced in time by yet another fad. Contrary views will emerge, as in *The Loyalty Effect: The Hidden Force Behind Growth, Profits, and Lasting Value* (3) by Frederick F. Reichheld, with Thomas Teal. The authors suggest that companies that care about long-term growth and profits should reconsider what they are doing to loyalty. Reichheld is a director of Bain & Company, a consulting firm in Boston, and the leader of a practice through which the firm provides a consultancy to clients on developing and maintaining loyalty. He writes:

> Firms that earned superior levels of customer loyalty and retention also earned consistently higher profits—and they grew faster as well.
>
> Loyalty as we conceive it is critically important as a measure of value creation and as a source of growth and profit.

Reichheld says that there is plenty of evidence that something about the current business paradigm is wrong. His solution is a return to loyalty-based management.

Assuredly, other solutions will be proposed by other consultants.

Management Fads

A management fad is a new scheme, a panacea, proposed by consultants or educators that is to provide a quick fix to solve management problems. Often the fad is the subject of a best-selling book.

With the new management scheme in place, employees will be happier, with improved morale and loyalty. Productivity and quality will surge. Growth and margins will be better than ever. And, the consultants say, managers will embrace the new ideas with enthusiasm since their advantages will be apparent and the new scheme will be perceived by them as making their jobs easier.

Managements seem to have a never-ending passion for a panacea to resolve their insecurity with whatever management practices are in place. And the consultants and educators accommodate their passions.

Management fads last but a few years, for only as long as the executives in charge continue to have an interest in them. For each of the management fads that have surfaced in the last half of this century, there has been an initial burst of excitement, a broad adoption of the fad by many organizations, and then, a quiet fading away. We do what the fads require, only to find later that what we were doing has become unimportant. But, a few fads will leave some good behind.

Seldom does management determine the impact that the new management practices will have on the organization's culture and plan well for their integration into the culture. Thus, the new fad creates anxiety and fear. Safety professionals cannot help but be caught up in the new fads. If their own practices have a sound theoretical and practical base, it will be obvious to them that quick fixes will not achieve what needs to be done.

Within the practice of safety, it's probable that, in time, what will be identified as another management fad has become very popular. Behavior modification and a psychological approach to safety are the vogue in some places. A report on a survey published in the December 1996 issue of *Industrial Safety & Hygiene*

News (4) indicates that although "behaviorism is booming," the typical backlash to management fads has begun. Consider this quotation:

> You don't have to interview many EHS [Environmental Health and Safety] experts to hear sniping about behavioral safety. Among the complaints: it blames accidents on employees and ignores management's responsibility for a safe operating system. It's costly and time-consuming to implement. It's tough to sustain employee involvement over the long haul. Proponents are more marketers than teachers. Don't be surprised if the criticism gets louder as the popularity of behavior safety grows.

The critics are right when they say that behavior modification and psychology of safety approaches inappropriately focus on employee performance, rather than on management's responsibility to provide a safe operating system. Also, whether the subject is productivity, cost-efficiency, quality, or safety, the methods applied to achieve acceptable employee behavior, and the psychology of employee supervision, are the same.

In an economic downturn, I would not want to be a safety professional who was principally perceived as fostering the idea that safety needs to be treated separately from the business process. To be perceived as an effective member of the management team and as supportive of achieving entity goals, a safety professional must keep in balance the design and engineering aspects and the management, operations, and task performance aspects of safety.

A Bit of History on Management Fads

With a near absolute certainty, it can be said that the never-ending parade of management fads will continue. Many of the management fads that became popular for a while in the second half of this century had their beginnings in Douglas McGregor's Theory X and Theory Y concepts. As new management fads arise, safety professionals may want to reflect on this background.

McGregor's book *The Human Side of Enterprise* was published in 1960. His view was that the leadership approach managers took was based on assumptions they made about the people they managed. McGregor grouped possible assumptions in what he called Theory X and Theory Y categories (5).

Theory X Assumptions

1. People do not like work and try to avoid it.
2. People do not like work, so managers have to control, direct, coerce, and threaten employees to get them to work toward organizational goals.
3. People prefer to be directed, to avoid responsibility; they want security; they have little ambition.

Theory Y Assumptions

1. People do not naturally dislike work; work is a natural part of their lives.
2. People are internally motivated to reach objectives to which they are committed.
3. People are committed to goals to the degree that they receive personal rewards when they reach their objectives.
4. People will both seek and accept responsibility under favorable conditions.
5. People have the capacity to be innovative in solving organizational problems.
6. People are bright, but under most organizational conditions their potentials are underutilized.

McGregor's view was that if a manager believed that workers had Theory X characteristics, they would be under close direction and control. But, McGregor believed that Theory Y characteristics prevail. His premise is that people can be self-actualizing, exercise intellectual capacities, and more fully utilize their capabilities—if they are given opportunities to participate in establishing goals and commit to them, and to influence the nature of their work.

In the introduction to a 1967 book by McGregor, *The Professional Manager* (6), Edgar H. Schein made these comments about *The Human Side of Enterprise*:

> . . . Doug's book has been seen as a plea for an attitude, a set of values toward people, symbolized by the term Theory Y. The essence of this attitude is to trust people, to grant them power to motivate and control themselves, to believe in their capacity to integrate their own personal values with the goals of the organization. Doug believed that individual needs can and should be integrated with organizational goals. In the extreme, Theory Y has meant democratic processes in management, giving people a greater voice in the making of decisions and trusting them to contribute rationally and loyally without surrounding them with elaborate control structures.

Concepts of employee empowerment and the many forms of participative management that have been tried over the years are reflections of McGregor's Theory Y.

How many management fads have American business applied and discarded in the past forty years? Here are the ones with which I am familiar: Management by Committee; Management by Consensus; Centralization and Decentralization; Quantitative Management; Pay for Performance; Intrapreneuring, Management by Walking Around; Theory Z; Transactional Analysis; Zero Defects; Quality Circles; The One-Minute Manager (the biggest rip-off I ever experienced in buying books); Matrix Management, Excellence (as in the book *In Search of Excellence);* the Managerial Grid; Employee Empowerment; and, in some companies, Total Quality Management.

Reengineering the Corporation (7) was written by James Champy and Michael Hammer and published in 1993. And since then, reengineering has been the fad in process. The authors proposed a method of reorganizing business around "processes" rather than around the traditional functions of marketing, selling, et cetera.

John Micklethwait and Adrian Wooldridge say in *The Witch Doctors: Making Sense of the Management Gurus* (8), that the reengineering practices proposed by Champy and Hammer "would work if people were unthinking automatons, without hearts and souls." They also say that, as is the case with most management fads, reengineering "is good at some things and not so good at others."

There are signs of a backlash: even Champy and Hammer have admitted that their reengineering concepts have fallen short of their goals.

Knowing that the parade of management fads will go on and on, safety professionals serve their own interests well if their professional practices are soundly based, and—always—if their work is perceived to be in support of management goals.

How We Are Perceived

Because of the turmoil in the business climate, it would be prudent for us to explore how we are perceived by those who have management responsibility for operations. Arthur D. Little issued a news release titled "Green Wall Between Companies' Environmental and Business Staffs Creates Major Roadblock to Successful Environmental Management" (9). It pertained to a survey in which this question was asked of managers of environmental, health, and safety: Within companies, what is the greatest impediment to integrating environmental, health, and safety into business? More than seven out of ten responded that there was either a lack of acceptance of environmental, health, and safety by the business staff, or that a culture of separateness had been created by the EH&S staff. Further, in a published review by human resources directors of what activity could readily be outsourced (10), opposite "health and safety," the indication is "no value added."

This problem of perception—the problem of others believing that we have created a culture of separateness, and of not providing value—must be taken seriously. This matter of perception requires that we assess what our goals should be.

Goals to Be Achieved by Safety Professionals

For some, what I propose here is substantially different from their perception of their roles. I believe that safety professionals should aspire to be perceived as:

- participants in effectively and economically applying available and limited resources to eliminate or reduce risk so that the greatest possible good to employees, employers, and society is obtained from those endeavors;

- contributing effectively to productivity, cost efficiency, and quality management, in addition to safety;
- active participants in achieving management goals.

I will refer often to being effective. If we say that we are professionals and that what we do is effective, we imply that our knowledge base is sound, and that the measures we propose to reduce risk are properly directed.

Being Participants in Achieving Management Goals

To establish what I believe are the paramount requirements of those with responsibility for operations, those with whom we are to participate in meeting their goals, I will refer to two highly respected writers—Peter F. Drucker and W. Edwards Deming. Both of these authors have had a major influence on my understanding of management concepts and practices.

For knowledge of sound business practices that will stand the test of time, I highly recommend the writings of Peter Drucker, not just those referenced here. This is what is said of Drucker in *The Witch Doctors: Making Sense of the Management Gurus* (8) in the chapter titled "Peter Drucker: The Guru's Guru."

> In the world of management gurus, however, there is no debate. Peter Drucker is undisputed alpha male. He is also one of the few thinkers from any discipline who can claim to have changed the world: he is the inventor of privatization, the apostle of a new class of knowledge workers, the champion of management as a serious intellectual discipline.
>
> . . . Drucker is the one management theorist who every reasonably well-educated person, however contemptuous of business or infuriated by jargon, really ought to read.

Drucker, in his book *Post-Capitalist Society* (11), said this:

> Economic performance is the first responsibility of a business. [Without economic performance] a business cannot discharge any other responsibilities, cannot be a good employer, a good citizen, a good neighbor.

What I have quoted from Drucker is in the same class as the first point in W. Edward Deming's often quoted "Condensation of the 14 Points of Management," as listed in his book *Out of the Crisis* (12). W. Edwards Deming is world-renowned with respect to quality management. The first of his fourteen management premises is that an entity should establish:

> [C]onstancy of purpose toward improvement of product and service, with the aim to become competitive and to stay in business, and to provide jobs.

We safety professionals must understand and function within the charge given to management to achieve economic goals, be competitive, stay in business, and provide jobs. If economic performance goals are not met and an organization does not stay in business, obviously, its safety professionals won't be needed.

American business is operating in a harsher climate with leaner staffs—a reality that requires more effective management. In that context, those who develop multiple skills, those who are problem solvers, and those who produce results surely will be perceived as more valuable.

In the real world of decision-making, for entities of every description, it must be recognized that some risks are more significant than others; resources are always limited; and the greatest good to employees, employers, and society is attained if resources are effectively applied to avoid, eliminate, or control hazards, and the risks that derive from them.

Safety professionals have a participatory obligation concerning the effective utilization by management of the resources available. Since resources are always limited, and since some risks are more significant than others, safety professionals must be capable of distinguishing the more important from the less important, identifying the potentials for the greatest harm or damage, and ranking risks in priority order.

We must also understand that business executives are risk takers who, appropriately, may not perceive that what we safety professionals propose must be given the highest priority at a given time within the spectrum of all the operating decisions they must consider.

But, in reality, we safety professionals are also risk takers. It is not possible to attain a risk-free environment, even in the most desirable situations. Setting a goal to achieve zero risk may seem laudable, but it requires chasing a myth. We must understand that even after the best practical methods are applied to what we propose, there will always be residual risk. And we are participants in accepting those risks.

Demands imposed by a highly competitive economy for more effective management performance are not going to go away. They compel safety professionals to seek to be perceived as problem solvers whose counsel is sought because of their successes as participants in achieving management goals.

Drucker on a Significant Societal Transition

Peter Drucker also stated, in *Post-Capitalist Society* (11), that the world is in transition to a knowledge society, that the primary resource in the post-capitalist society will be knowledge, and that the leading social group will be knowledge workers. He writes:

> In the post-capitalist society, it is safe to assume that anyone with knowledge will have to acquire new knowledge every four or five years, or become obsolete.

In *Managing in a Time of Great Change* (1), Drucker predicts that by the end of the century, knowledge workers will amount to a third or more of the workforce. He also says that knowledge is the only meaningful resource today, and that continuing to expand learning is critical to knowledge workers.

All that Drucker says applies to the practice of safety. Very few safety professionals would say that they are not knowledge workers or that the content of their work does not continuously change. Since transitions in the practice of safety will surely take place, it follows that our knowledge base needs a constant examination as to its applicability and effectiveness. Obviously, safety professionals must endlessly acquire new knowledge and skills to maintain a proper level of professional practice.

If the application of our knowledge is not effective, does not prove itself in action, and does not produce results, can we expect, realistically, that what we do will be considered valuable?

Maintaining a Solid Career Base

There has been a proliferation of business writings advising professionals that they cannot design their lives around organizations that may not exist in the future, that loyalty is a thing of the past, that one can expect to have several employers, and that knowledge and skill needs grow exponentially. Writers then proceed to give advice on how to maintain a solid career base.

Dr. Roger L. Brauer, executive director of the Board of Certified Safety Professionals, gave a presentation titled "Challenges and Opportunities" at the 1996 Professional Development Conference held by the American Society of Safety Engineers. Since I thought his presentation was sound, I am summarizing his outline and adding my own views.

To maintain a solid career base, a professional should:

- Have a plan that considers family, community, knowledge of capabilities and limitations, and professional career desires—and visit that plan regularly.
- Never stop learning, develop a solid professional knowledge and skill base, anticipate changes and creatively develop the new solutions they require, learn by doing and benefit from the mistakes that will be made, and, above all, strive for excellence. Having a master's degree has become a more frequently cited employment desirable.
- Have a good understanding of sound management practices, knowing that many management fads will be experienced during one's career.
- Be active in professional organizations for an exchange of knowledge concerning new developments in technology and for the networking opportunities they provide.

- Take care of him/herself—physically, mentally, and emotionally.
- Obtain professional credentials—particularly the Certified Safety Professional designation, which has grown in significance as an employment criteria.
- Be alert to and be prepared for opportunities, knowing that career changes will take place.

It should be understood by all professionals that they have to take responsibility for their own employment security, and that it would be folly to program their lives within an organization that may exist only temporarily. Also, they must realize that their continued employment potential is being determined by whether their skills and capabilities are viewed by their employers as supportive in achieving entity goals. Transitions in recent years in the technical requirements for the practice of safety require that safety professionals expand their skills. Doing so results in their having additional opportunities to be perceived as more valued employees.

IMPACT OF APPLIED ERGONOMICS

Applied ergonomics became a major element in the practice of safety as the significance of ergonomics-related injuries and illnesses was recognized within the entire spectrum of incidents resulting in harm to employees. As to that relationship, I cite an Aetna study (13) indicating that about 50 percent of work injuries and illnesses and 60 percent of total workers compensation costs are ergonomics related. Comparable figures have appeared in several writings.

Ergonomics developed out of the need to make work less stressful and more comfortable. And these are laudable goals that can be more effectively achieved if managements understand that the application of ergonomics principles also serves their productivity and cost-efficiency goals. Case histories are now prevalent describing ergonomics applications that were initiated to resolve injury and illness problems, but also achieved improvements in productivity and cost-efficiency.

Emphasizing the possible productivity and cost-efficiency gains from applied ergonomics is a somewhat different approach from that found in the prominent ergonomics literature. This novel idea gets management attention.

I had previously defined ergonomics as the art and science of designing the work to fit the worker. I now use this definition:

> Ergonomics is the art and science of designing the work to fit the worker—to achieve optimum productivity and cost efficiency, and minimum risk of injury.

Safety professionals who take the approach implied by this definition have opportunities for additional accomplishment and recognition. Ergonomics is design-based, as is the entirety of the practice of safety.

There is a developing recognition of the benefits to be obtained by scoping ergonomics designs to encompass productivity, cost-efficiency, quality, and safety as parts of an interrelated whole. Because it gives a heavy emphasis to such benefits, a workshop in the program at The College of Engineering at the University of Wisconsin at Madison is unique. A brochure titled Advanced Ergonomics Application Workshop says that participants will learn, among other things, to develop job design guidelines to improve productivity and quality; prioritize job design factors; optimize and balance good ergonomics and productivity for material handling tasks; and determine the impact on productivity through a before-and-after analysis.

Relating ergonomics to safety, productivity, and quality as an integrated whole spells opportunity for safety professionals who seek to be perceived as supporting entity goals and who want to build solid job security.

About Quality Management

A review of the considerable literature on quality management leads to this conclusion: for paragraph after paragraph in most texts and papers, the word "quality" can be replaced with "safety," and the premise remains sound. There is a remarkable kinship between sound quality management principles and sound hazards management principles. Many safety professionals are now engaged in quality management initiatives. Some say that they derive greater satisfaction from their work, that their status within their organizations has improved, and that they are viewed as more effective members of the management team.

Methods to achieve product quality are identical in most respects with the methods used to achieve superior results in safety. Also, the processes addressed in a continuous improvement initiative to minimize product defects are the same processes in which injuries, illnesses, and environmental incidents occur. Is the practice of safety, as suggested by Dr. Thomas A. Selders, but another means of achieving quality performance?

Safety professionals engaged in sound quality management initiatives think differently about the root causal factors for incidents that may result in injury or damage. They become indoctrinated in the concept of continuous improvement of processes to improve quality, and safety. There is an evident moving away from this principle, presented by Heinrich in *Industrial Accident Prevention* (14), where he writes that "methods of control must be directed toward man failure."

Where does this new knowledge, this new emphasis take them? A thought process evolves whereby, when a risk situation arises, they pose questions concerning the system, rather than focusing principally on the behavior of individuals. That leads to a more realistic identification and analysis of hazards, which is vital to successful safety practice.

W. Edwards Deming has had a great influence on quality management throughout the world. I admit to being much impressed by his work. In *Out of the Crisis* (12) he stated his belief that a large majority of operational problems are systemic and can be resolved only by management, and that responsibility for only a relatively small remainder lies with the worker.

I believe that Deming's premise is valid. Serious thought needs to be given to it by safety professionals since it applies to the practice of safety. Deming's emphasis on constantly improving the system rather than expecting employees to do what they cannot do poses questions that safety professionals should consider concerning the content and effectiveness of their work. Out of that thinking should come an awareness of opportunities.

Combined Responsibility for Environmental Affairs, Health, and Safety

In the report on a survey published by *Industrial Safety & Hygiene News* in December 1996 (4), it is said that "About one-third of facilities surveyed operate combined EHS departments at the highest level of the organization." That arrangement allows the development of one objective strategy and one source of advice. It makes no sense, from economic or effectiveness viewpoints, to look at only one aspect of a situation when there is an overlapping of safety, health, and environmental risks.

Whatever the professional history of those now having expanded responsibilities for environmental affairs, safety, and health, an urgency has developed for additional technical knowledge and skills, and for more effective capabilities in people relations and communications.

As a matter of career development and job security, safety professionals should consider reaching out to take environmental affairs into their responsibilities. That spells opportunity.

Conclusion

Transitions will surely continue in business practices and in the practice of safety. To the astute practitioner, these transitions can be perceived as opportunities if they develop and maintain a solid career base.

A phrase that is just emerging but which will be heard more often spells additional opportunity for accomplishment and recognition by enterprising safety professionals. That phrase is: avoid bringing hazards into the workplace.

REFERENCES

1. Drucker, Peter F. *Managing In A Time Of Great Change.* New York: Truman Talley Books/Dutton, 1995.
2. Heckscher, Charles. *White-Collar Blues.* New York: BasicBooks, 1995.
3. Reichheld, Frederick, with Thomas Teal. *The Loyalty Effect: The Hidden Force Behind Growth, Profits and Lasting Values.* Boston: Harvard Business School Press, 1996.
4. *Industrial Safety & Hygiene News,* December, 1996.
5. McGregor, Douglas. *The Human Side of Enterprise.* New York: McGraw-Hill, 1960.
6. McGregor, Douglas. *The Professional Manager.* New York: McGraw-Hill, 1967.
7. Champy, James, and Michael Hammer. *Reengineering the Corporation.* London: Nicholas Brealey, 1993.
8. Micklethwait, John, and Adrian Wooldridge. *The Witch Doctors: Making Sense of the Management Gurus.* New York: Times Books/Random House, 1996.
9. Little, Arthur D. Undated News release. "Green Wall Between Companies' Environmental and Business Staffs Creates Major Roadblock to Successful Environmental Management," Cambridge, MA: Arthur D. Little Inc.
10. *Vision of the Future: Role of Human Resources in the New Corporate Headquarters.* Corporate Leadership Council, the Advisory Board Company, 1995.
11. Drucker, Peter F. *Post-Capitalist Society.* New York: HarperCollins Publishers, 1993.
12. Deming, W. Edwards. *Out of the Crisis.* Cambridge, MA: Center for Advanced Engineering Study, Massachusetts Institute of Technology, 1986.
13. Ergonomics Workshop. Aetna Life and Casualty, Hartford, CT, 1990.
14. Heinrich, H. W. *Industrial Accident Prevention.* 3rd ed. New York: McGraw-Hill, 1950.

Chapter 2

Principles for the Practice of Safety: A Basis for Discussion

Introduction

For the practice of safety to be recognized as a profession, it must have a sound theoretical and practical base, the application of which will be effective in attaining hazard avoidance, elimination, or control and, thereby, achieving a state in which the risks deriving from those hazards are at an acceptable level. My belief is that there is a generic base for the work of safety professionals that must be understood and applied if we are to be effective.

But, safety professionals have not yet agreed on those fundamentals. We take a variety of approaches to achieving safety, each based on substantively different premises. They can't all be right or equally effective. To promote a discussion toward establishing a sound theoretical and practical base for the practice of safety, I offer a list of general principles, statements, and definitions that I believe to be sound. The list is a beginning: It is not complete.

I hope that this list will encourage dialogue by those who have an interest in moving the state of the art forward.

A. Hazards Are the Generic Base of and the Justification for the Existence of the Entirety of the Practice of Safety

1. Hazards include any aspect of technology or activity that produces risk. (Fischhoff, 1)
2. A hazard is defined as the potential for harm: Hazards include the characteristics of things and the actions or inactions of people.

3. Hazards are the generic base of and the justification for the existence of the entirety of the practice of safety. If there were no hazards, safety professionals need not exist.
4. If a hazard is not avoided, eliminated, or controlled, its potential will be realized, and a hazards-related incident will occur that may or may not result in harm or damage, depending on exposures.
5. A hazards-related incident, a HAZRIN, is an unplanned, unexpected process of multiple and interacting events, deriving from the realization of uncontrolled hazards and occurring in sequence or in parallel, which is likely to result in harm or damage.
6. Hazards-related incidents, even the ordinary and frequent, are complex and have multiple and interacting causal factors.
7. Hazard analysis is the most important safety process in that, if that fails, all other processes are likely to be ineffective. (Johnson, 2)
8. If hazard identification and analysis do not relate to actual causal factors, corrective actions will be misdirected and ineffective.

B. Defining Risk and Safety, and Their Relationship

1. Risk is defined as a measure of the probability of a hazards-related incident occurring, and the severity of harm or damage that could result.
2. Safety is defined as that state for which the risks are judged to be acceptable.
3. All risks to which the practice of safety applies derive from hazards: there are no exceptions.
4. Setting a goal to achieve a zero-risk environment may seem laudable, but doing so requires chasing a myth.
5. It is impossible to attain a risk-free environment. Even in the most desirable situations, there will still be residual risk after application of the best, most practical methods.
6. For an operation to proceed, its risks must be acceptable.

C. Defining the Practice of Safety

1. The entirety of purpose of those accountable for safety, in fulfilling their societal responsibilities, is to manage their endeavors with respect to hazards, so that the risks deriving from those hazards are at an acceptable level.
2. The practice of safety:
   - serves the societal need to prevent or mitigate harm or damage to people, property, and the environment, deriving from hazards;
   - is based on knowledge and skill as respects Applied Engineering, Applied Sciences, Applied Management, and Legal/Regulatory and Professional Affairs;

- is accomplished through
  - anticipating, identifying, and evaluating hazards, and
  - actions taken to avoid, eliminate, or control those hazards
- has as its ultimate purpose attaining a state for which the risks are judged to be acceptable.

3. While safety professionals may undertake many tasks in their work, the underlying purpose of each task is to have the attendant risks at an acceptable level. Every element of a safety initiative should relate to hazards and the risks that derive from them.
4. Professional safety practice requires consideration of two distinct aspects of risk:
   - avoiding, eliminating, or reducing the *probability* of a hazards-related incident occurring; and
   - minimizing the *severity* of harm or damage, if an incident occurs.
5. There are three major elements in the practice of safety:
   - in the design processes, pre-operational—where the opportunities are greatest for identifying and analyzing hazards and for their avoidance, elimination, or control;
   - in the operational mode—where, integrated within a continuous improvement process, hazards are identified, evaluated, eliminated, or controlled, *before* their potentials are realized and hazards-related incidents occur; and
   - post incident—through investigation of hazards-related incidents to determine and eliminate or control their causal factors.

D. On Achieving the Theoretical Ideal for Safety

1. The theoretical ideal for safety is achieved when all risks deriving from hazards are at an acceptable level.
2. That definition serves, generally, as a mission statement for the work of safety professionals, and as a reference against which each of the many activities in which they engage can be measured.
3. A statement in *Why TQM Fails and What To Do About It,* by Graham Mark Brown, Darcy E. Hitchcock, and Marsha L. Willard, provides, with minimum modification, a basis for review to determine how near operations are to achieving the theoretical ideal for safety. In the following quotation, the word "safety" appears twice. In the first instance, it replaces TQM; in the second, it replaces "quality": "When safety is seamlessly integrated into the way an organization operates on a daily basis, safety becomes not a separate activity for committees and teams, but the way every employee performs job responsibilities." (3)
4. That statement implies that a system of expected behavior is in place as a reflection of an organization's culture that requires a superior level of safety performance.

5. When safety is seamlessly integrated into the way an organization functions on a daily basis, a separately identified safety program is not needed since all actions required to achieve safety would be blended into operations.
6. Thus, the theoretical ideal for a safety program, is nothing. (Program: a plan or schedule to be followed.)

E. On Organizational Culture

1. An organization's culture determines the level of safety to be attained. What the board of directors or senior management decides is acceptable for the prevention and control of hazards is a reflection of its culture.
2. An organization's culture consists of its values, beliefs, legends, rituals, mission, goals, and performance measures, and its sense of responsibility to its employees, its customers, and its community, all of which are translated into a system of expected behavior.
3. Management obtains, as a derivation of its culture, as an extension of its system of expected behavior, the hazards-related incident experience that it establishes as tolerable. For personnel in the organization, "tolerable" is their interpretation of what management does.
4. An organization's culture, translated into a system of expected behavior, determines management's commitment or noncommitment to safety; involvement and direction; accountability system; safety policy; safety organization; standards for workplace and work methods design; requirements for continuous improvement; and the behavioral climate that is to prevail concerning management and personnel factors (leadership, training, communication, adherence to safe work practices, et cetera).
5. Management commitment is questionable if the accountability system does not include safety performance measures that impact on the well-being of those responsible for results.
6. What management does, rather than what management says, defines the actuality of commitment or noncommitment to safety.
7. Principal evidence of an organization's culture with respect to hazards management is demonstrated through the design decisions that determine what the facilities, hardware, equipment, tooling, materials, configuration and layout, the work environment, and the work methods are to be.
8. If the design of a system (the facilities, equipment, work methods, etc.) does not achieve minimum risk, superior results with respect to safety cannot be attained, even if personnel and management factors approach the ideal.
9. Where the culture (the system of expected behavior) demands superior safety performance, the design and engineering aspects of safety, and the management, operations, and task performance aspects of safety are well-balanced.
10. Major improvements in safety will be achieved only if a culture change takes place—only if major changes occur in the system of expected behavior.

F. Concerning Leadership, Training, and Behavior Modification

1. Effective leadership, training, communication, persuasion, behavior modification, and discipline are vital aspects of hazards management, without which superior results cannot be achieved.
2. But training and behavior modification, et cetera, are often erroneously applied as solutions to problems, with unrealistic expectations. Such personnel actions have limited effectiveness when incident causal factors derive from workplace and work methods design decisions. (It is recognized that in certain situations personnel actions are the only preventive measures that can be taken.)
3. Earl D. Heath and Ted Ferry write in *Training in the Workplace: Strategies for Improved Safety and Performance*: "Employers should not look to training as the primary method for preventing workplace incidents that result in death, injury, illness, property damage or other down grading incidents. They should see if engineering revisions can eliminate the physical safety and health hazards entirely." (4)
3. As an idea, the substance of, but not the precise numbers of, what is being called "Deming's 85-15 Rule" applies to all aspects of the practice of safety. The following is from *The Deming Management Method* by Mary Walton: "Deming's 85-15 Rule . . . holds that 85 percent of the problems in any operation are within the system and are the responsibility of management, while only 15 percent lie with the worker." (5)
4. In *Out of the Crisis* by W. Edwards Deming, this is how the subject is treated: "I should estimate in my experience most troubles and most possibilities for improvement add up to proportions something like this: 94% belong to the system (responsibility of management); 6% special." (6)
5. The premise is valid: that a large majority of the problems in any operation are systemic, deriving from the workplace and the work methods created by management, and can be resolved only by management; responsibility for only the relatively small remainder lies with the worker.
6. Extrapolating from Deming, a large majority of the causal factors for hazards-related incidents will be systemic, and a small minority will be principally employee focused.
7. Problems that are in the system can only be corrected by a redesign of the system. If system design and work methods design are the problems, the capability of employees to help is principally that of problem identification.
8. This is from *Out of the Crisis,* in which Deming, referencing Joseph M. Juran, speaks of workers being "handicapped by the system": "The supposition is prevalent the world over that there would be no problems in production or in service if only our production workers would do their jobs in the way that they were taught. Pleasant dreams. The workers are handicapped by the system, and the system belongs to management. It

was Dr. Joseph M. Juran who pointed out long ago that most of the possibilities for improvement lie in action on the system, and that contributions of production workers are severely limited." (6)

9. While employees should be trained and empowered up to their capabilities and encouraged to make contributions to safety, they should not be expected to do what they cannot do.
10. While safety is a line responsibility, it should be understood that achievements by management at an operating level are limited by the previously made workplace and work methods design decisions. If the design of the system presents excessive operational risks for which the cost of retrofitting is prohibitive, administrative controls, perhaps the only actions that can be taken, will achieve less than superior results.

G. Significance of Design and Engineering

1. Deming got it right: a large majority of the problems in an operation are systemic, deriving from the workplace and work methods created by management, and responsibility for only the relatively small remainder lies with the workers.
2. Then the greatest strides forward with respect to safety, health, and the environment will be made through the design and redesign processes.
3. For the practice of safety, the term design processes applies to:
   - facilities, hardware, equipment, tooling, selection of materials, operations layout, and configuration; and
   - work methods and procedures, personnel selection standards, training content, management of change procedures, maintenance requirements, and personal protective equipment needs, et cetera.
4. Design and engineering applications that determine the workplace and work methods are the preferred measures of prevention since they are more effective in avoiding, eliminating, and controlling risks.
5. Over time, the level of safety achieved will relate directly to the caliber of the initial design of the workplace and work methods, and their subsequent redesign in a continuous improvement endeavor.
6. A fundamental design goal is to have processes that are error proof. Juran and Gryna, in *Quality Planning and Analysis,* speak appropriately of "Error Proofing the Process," in these quotations: "An important element of prevention is the concept of designing the process to be error free through "error proofing" (the Japanese call it pokayoke or bakayoke). A widely used form of error proofing is the design (or redesign) of the machines and tools (the "hardware") so as to make human error improbable or even impossible." (7)
7. Requirements to achieve an acceptable risk level in the design process can usually be met without great cost if the decision making takes place

sufficiently upstream. When that does not occur, and retrofitting to eliminate or control hazards is proposed, the cost may be so great as to be prohibitive.

H. On System Safety

1. In the *Scope and Functions of the Professional Safety Position* (8) issued by the American Society of Safety Engineers, it is said that the safety professional is to anticipate, identify, and evaluate hazardous conditions and practices, and develop hazard control designs, methods, procedures, and programs. Those are valid statements.
2. If safety professionals are to anticipate hazards, they must participate in the design process. To be involved in the design process effectively, they must be skilled in hazard analysis and risk-assessment techniques. Being a participant in the design process and using hazard analysis and risk-assessment techniques are the basics of system safety.
3. Applied system safety requires a conscientious, planned, disciplined, and systematic use of special engineering and management tools *on an anticipatory and forward-looking basis.*
4. R. L. Browning's premise, as stated in *The Loss Rate Concept in Safety Engineering,* is sound: "As every loss event results from the interactions of elements in a system, it follows that all safety is system safety." (9)
5. A significant premise of system safety is that hazards are most effectively and economically anticipated, avoided, or controlled in the initial design process.
6. For both the workplace design aspects of safety, and the management, operations, and task performance aspects of safety, application of hazard analysis and risk-assessment methods is vital to achieving acceptable risk level.
7. Joe Stephenson, in *System Safety 2000,* expressed this view, which is sensible: "The safety of an operation is determined long before the people, procedures, and plant and hardware come together at the work site to perform a given task." (10)

I. Significance of Ergonomics

1. Ergonomics is the art and science of designing the work to fit the worker—to achieve optimum productivity and cost efficiency, and minimum risk of injury.
2. Estimates frequently appear indicating that about 50 percent of work injuries and illnesses and 60 percent of total claims costs are ergonomics related.

3. Assume that estimate to be close to reality. Then, to be effective, a safety generalist giving counsel for workplace safety and health must be deeply engaged in ergonomics.
4. Effective application of ergonomics principles requires that workplace and work methods analyses properly identify the possible stresses to employees (the hazards) and that they be addressed in the design or redesign of the workplace and work methods.
5. Ergonomics developed out of the need to make work less stressful and more comfortable. Those are laudable goals, which can be more effectively achieved if managements are convinced that the application of ergonomics principles also serves their productivity and cost-efficiency goals.
6. It should be the exception when an ergonomics analysis is made only for safety purposes. Rather, the analysis should address the aspects of productivity, cost efficiency, and risk reduction, all in one study.

J. Setting Priorities and Utilizing Resources Effectively

1. These principles are postulated:
   - All hazards do not present equal potential for harm or damage.
   - All incidents that may result in injury, illness, or damage do not have equal probability of occurrence, nor will their adverse outcomes be equal.
   - Some risks are more significant than others.
   - Resources are always limited. Staffing and money are never adequate to attend to all risks.
   - The greatest good to employees, employers, and society is attained if available resources are effectively and economically applied to avoid, eliminate, or control hazards and the risks that derive from them.
2. Since resources are always limited, and since some risks are more significant than others, safety professionals must be capable of distinguishing the more important from the less important.
3. Professional safety practice requires that the potentials for the greatest harm or damage be identified for the decision makers, and that a ranking system be applied to proposals made to avoid, eliminate, or control hazards.
4. Safety professionals must, therefore, be capable of using formalized hazard analysis and risk-assessment methods.
5. Causal factors for low-frequency incidents resulting in severe harm or damage may be different from the causal factors for more frequently occurring incidents. Such low-frequency incidents often involve unusual or non-routine work, high energy sources, nonproductive activities, and certain construction situations. (Petersen, 11)
6. Thus, safety professionals must undertake a separate and distinct activity to seek those hazards that present the most severe injury or damage potential, so that they can be given priority consideration.

K. On Incident Causation

1. For almost all hazards-related incidents, even those that seem to present the least complexity, there are multiple causal factors that derive from *less than adequate* workplace and work methods design and management, operations, and personnel task performance practices.
2. In *MORT Safety Assurance Systems,* William G. Johnson wrote succinctly about the multifactorial aspect of incident causation: "Accidents are usually multifactorial and develop through relatively lengthy sequences of changes and errors. Even in a relatively well-controlled work environment, the most serious events involve numerous error and change sequences, in series and parallel." (2)
3. In the hazards-related incident process, deriving from those multiple causal factors:
   - there are unwanted energy flows or exposures to harmful environments (Haddon, 12,13);
   - a person or thing, or both, in the system is stressed beyond the limits of tolerance or recoverability (McClay, 14);
   - the incident process begins with an initiating event in a series of events;
   - multiple interacting events occur, sequentially or in parallel, over time and influencing each other to a conclusion that may or may not result in injury or damage (Benner, 15)
4. *Severity potential* should determine whether hazards-related incidents should be considered significant, even though serious harm or damage did not occur.
5. After a review of theorems developed in the 1920s, H. W. Heinrich presented, in *Industrial Accident Prevention,* a causation model which was illustrated by a "domino sequence." Heinrich wrote: "From this sequence of steps in the occurrence of accidental injury, it is apparent that man failure is the heart of the problem. Equally apparent is the conclusion that methods of control must be directed toward man failure." (16)
6. This is an expression from Johnson in *MORT Safety Assurance Systems:* "Safety texts have used a row of falling dominoes as a model illustrating the accident sequence. This may be such a gross simplification as to limit understanding." (2)
7. Heinrich's causation model has been prominently used by safety practitioners. Other causation models are extensions of it. Often, the wrong advice is given when causation models and incident analysis systems focus primarily and almost exclusively on:
   - characteristics of the individual
   - unsafe acts being the primary causes of workplace incidents; and
   - corrective measures being directed principally to correct "man failure," mainly to affect the behavior of an individual

8. Heinrich also wrote that "a total of 88 percent of all industrial accidents . . . are caused primarily by the unsafe acts of persons." (16)
9. Those who continue to promote the idea that 88 or 90 or 92 percent of all industrial accidents are caused primarily by the unsafe acts of persons do the world a disservice. Investigations that properly delve into causal factors indicate that premise to be invalid.
10. Johnson expressed his concern over the common use of the terms "unsafe conditions" and "unsafe acts" in this manner: "These categories of errors are extensively used in accident data collection. However, the fact remains that concepts of unsafe conditions and unsafe acts are usually simplistic and definitions are variable from place to place." (2)
11. This statement appears under "Performance Errors" in the 1994 publication titled *Guideline to Use of the Management Oversight and Risk Tree*: "It should be pointed out that the kinds of questions raised by MORT are directed at systemic and procedural problems. The experience shows there are few 'unsafe acts' in the sense of blameful work level employee failures. Assignment of 'unsafe act' responsibility to a work level employee should not be made unless or until the preventive steps of: (1) hazard analysis, (2) management or supervisory detection, and (3) procedures safety review have been shown to be adequate." (17)
12. Incident investigation should initially address the work system. That means replacing "Heinrichian" based premises that concentrate on "methods of control . . . directed toward man failure" with a concept that:
    - commences with inquiries to determine whether causal factors derive from workplace design decisions; and
    - promotes ascertaining whether the design of the work methods was overly stressful or error-provocative, or whether the immediate work situation encouraged riskier actions than the prescribed work methods.

L. Performance Measures

1. If the practice of safety professionals is based on sound science, engineering, and management principles, it follows that they should be able to provide measures of performance that reflect with some degree of accuracy the outcomes of the hazards management initiatives they propose.
2. Understanding the validity and shortcomings of our performance measures is an indication of the maturity of the practice of safety as a profession.
3. Safety professionals must understand that the quality of the management decisions made to avoid, eliminate, or control hazards are impacted directly by the validity of the information they provide through their performance measurement systems. Their ability to provide accurate information to be used in the decision-making is a measure of their effectiveness.

4. Since safety achievements in an organization are a direct reflection of its culture, and since it takes a long time to change a culture, short-term performance measures should be examined cautiously as to validity.
5. As the sample base, the number of hours worked, increases in size, the historical incident record has an increasing degree of confidence for incidents that have frequent, probable, occasional, and sometimes remote occurrence probabilities as:
   - a measure of the quality of safety in place; and
   - a general, but not hazard specific, predictor of the experience that will develop in the future.
6. But no statistical, historical performance measurement system can assess the quality of safety in place, to include risks for which the probability of occurrence is remote or improbable, and the severity of outcomes is critical or catastrophic since such events seldom appear in the statistical history. (Example: a risk assessment concludes that a defined catastrophic event has an occurrence probability of once in 200 plant operating years.)
7. Even for the large organization with significant annual hours worked, in addition to historical data, hazard-specific and qualitative performance measures (safety audits, The Critical Incident Technique) are also necessary, particularly to identify low-probability/severe-consequence risks.
8. Statistical process controls (cause-and-effect diagrams, control charts, et cetera), as applied in quality management, can serve as performance measures for safety, if they are used prudently and with caution.
9. Incidents resulting in severe injury or damage seldom occur and would rarely be included in the plottings on a statistical process control chart. Although such a chart may indicate that a system is in control, it could be deluding if it was presumed that the likelihood of low-probability incidents occurring that could result in severe harm or damage was encompassed in the plottings.

M. On Safety Audits

1. Safety audits must meet this definition to be effective: A safety audit is a structured approach to provide a precise evaluation of safety effectiveness, a diagnosis of safety problems, a description of where and when to expect trouble, and guidelines concerning what should be done about the problems.
2. The paramount goal of a safety audit is to influence favorably the organization's culture. Donald W. Kase and Kay J. Wiese concluded properly in *Safety Auditing: A Management Tool* that: "Success of a safety auditing program can only be measured in the terms of the change it effects on the overall culture of the operation, and enterprise that it audits." (18)
3. Since evidence of an organization's culture and its management commitment is first demonstrated through its upstream design and engineering

practices, safety audits that do not evaluate the design processes are incomplete and fall short of the definition of an audit.

4. Safety audits must also properly measure management commitment, primary evidence of which is a results-oriented accountability system. If such an accountability system does not exist, management commitment is questionable.

References

1. Fischhoff, Baruch. *Risk: A Guide to Controversy, Appendix C, Improving Risk Communication.* Washington, D.C.: National Academy Press, 1989.
2. Johnson, William G. *MORT Safety Assurance Systems.* New York: Marcel Dekker, 1980.
3. Brown, Graham Mark, Darcy E. Hitchcock, Marsha L. Willard. *Why TQM Fails and What to Do About It.* Burr Ridge, IL: Irwin Professional Publishing, 1994.
4. Heath, Earl D., and Ted Ferry. *Training in the Work Place: Strategies for Improved Safety and Performance.* Goshen, NY: Aloray, 1990.
5. Walton, Mary. *The Deming Management Method.* New York: Putnam, 1986.
6. Deming, W. Edwards. *Out of the Crisis.* Cambridge, MA: Center for Advanced Engineering Study, Massachusetts Institute of Technology, 1986.
7. Juran, J. M., and Frank M. Gryna. *Quality Planning and Analysis.* New York: McGraw-Hill, 1983.
8. *Scope and Functions of the Professional Safety Position.* Des Plaines, IL: American Society of Safety Engineers, 1993.
9. Browning, R. L. *The Loss Rate Concept in Safety Engineering.* New York: Marcel Dekker, 1980.
10. Stephenson, Joe. *System Safety 2000.* New York: Van Nostrand Reinhold, 1991.
11. Petersen, Dan. *Safety Management.* Goshen, NY: Aloray, Inc., 1988.
12. Haddon Jr., William J. *Preventive Medicine: The Prevention of Accidents.* Boston: Little, Brown, 1966.
13. Haddon Jr., William J. "On the Escape of Tigers: An Ecological Note," *Technology Review* (May 1970).
14. McClay, Robert E. "Toward a More Universal Model of Loss Incident Causation," *Professional Safety* (January/February, 1989).
15. Benner Jr., Ludwig. "Accident Investigation: Multilinear Sequencing Methods." *Journal of Safety Research* (June 1975).
16. Heinrich, H. W. *Industrial Accident Prevention.* 3rd ed. New York: McGraw-Hill, 1950.
17. Guide to Use of the Management Oversight and Risk Tree. U.S. Department of Energy, Office of Safety and Quality Assurance, SSDC–103. Washington, D.C.: November 1994.
18. Kase, Donald W., and Kay J. Wiese. *Safety Auditing: A Management Tool.* New York: Van Nostrand Reinhold, 1990.

Chapter 3

Defining the Practice of Safety

INTRODUCTION

After participating with safety practitioners in what he considered a baffling discussion of concepts, a highly regarded professor observed that what we who call ourselves safety professionals actually do will never be accepted as a profession by those outside our field until we agree on a clear definition of our practice. I agree with that premise, and will define the practice of safety in this essay.

It is a basic requirement of a profession to develop a precise and commonly accepted language that clearly presents an image of the profession. And the terminology used by safety professionals should also convey an immediate perception of their practice.

In his book *General Insurance* (1), David L. Bickelhaupt made a significant statement about the need for clear communications that speaks to the purpose of this treatise.

> Terminology becomes important in the serious study of any subject. It is the basis of communication and understanding. Terms that are loosely used in a general or colloquial sense can lead only to misunderstanding in a specialized study area such as insurance.

Similarly, clear terminology and avoiding terms that lead to misunderstanding is necessary in the practice of safety, which surely requires highly specialized study.

Defining Safety

We must agree on what we mean when we use the word *safety,* as in *the practice of safety.* If we cannot, how can we assume that we are communicating with each other or with those outside our profession when we use the term?

Dictionary definitions of safety are commonly given in safety literature. Since they are based on absolutes, such definitions are of little value to us. One dictionary defines *safety* as: "the quality of being safe; freedom from danger or injury." And *safe* is defined as: "free from or not liable to danger; involving no danger, risk or error."

Attaining a state in which there is no danger or risk that would qualify for dictionary definitions of safety is not possible. No environment can be absolutely safe.

In the *1989 Annual Report* (2) issued by the National Safety Council, safety was defined as "the control of hazards to attain an acceptable level of risk."

In their book *Introduction to Safety Engineering* (3), David S. Gloss and Miriam Gayle Wardle give this definition of safety: "Safety is the measure of the relative freedom from risks or dangers. Safety is the degree of freedom from risks and hazards in any environment." Also, in answering the question "How safe is safe?" they say "Safety is relative—nothing is 100% safe under all conditions."

In *Occupational Safety Management and Engineering* (4), Willie Hammer wrote this: "Safety—frequently defined as free from hazards. However, it is practically impossible to completely eliminate all hazards. Safety is therefore a matter of relative protection from exposure to hazards; the antonym to danger."

William W. Lowrance stated in *Of Acceptable Risk: Science and the Determination of Safety* (5) that: "We will define safety as a judgment of the acceptability of risk. . . . A thing is safe if its risks are judged to be acceptable."

Borrowing from all those whom I have quoted, this definition of safety is proposed as being applicable to the practice of safety in which all safety professionals are engaged. It is also credible in dealing with decision-makers:

> Safety is defined as that state for which the risks are judged to be acceptable.

Safety professionals must be aware that implementing all that they propose may reduce risk significantly, but that risk cannot be eliminated entirely. No matter how effective the preventive measures taken, there will always be a residual risk if an operation continues. It is unrealistic to assume that an environment could exist in which the probability is zero of an injurious or damaging event occurring.

Determining whether a thing, an activity, or an environment is safe requires making a judgmental decision. People make countless decisions to participate in activities for which they judge the risks to be tolerable.

Risk assessments made by scientists do not determine whether a thing is safe. Results of their studies will establish the probability of undesirable events occurring under given circumstances and the severity of their outcomes. Deciding that a thing is safe or not safe requires judgments of whether the probability of an undesired incident occurring and the severity of its outcome is acceptable.

In the definitions of safety previously quoted, the terms *risk* and *hazards* are used. In establishing what the practice of safety is all about, clear understanding of those two terms is also necessary.

Defining Risk

Arriving at a definition of risk applicable to the practice of safety that could be used convincingly in discussions with decision makers was not easy. Risk is a word that has too many meanings. Executives with whom safety professionals deal may hear the word used in several contexts in a given day.

Taking a business risk, a speculative risk, offers the possibility of gain or loss. That implies a meaning of risk different from that to which the practice of safety applies. Risks with which safety professionals are involved can only have adverse outcomes.

Definitions of risk in risk management and insurance literature were reviewed with the expectation that they would be helpful. But they do not meet the purposes of the practice of safety. A few examples follow.

In *Principles of Insurance* (6), R. I. Mehr and E. Commack write that "risk is uncertainty concerning loss." M. R. Greene defines risk in *Risk and Insurance* (7) as "uncertainty that exists as to the occurrence of some event." In Bickelhaupt's *General Insurance* (1), he writes: "Basically, risk is uncertainty, or lack of predictability." In *Risk Management and Insurance* (8), C. Arthur Williams and Richard M. Heins define risk as "the variations in the outcomes that could occur over a specified period in a given situation."

Risk fundamentally implies uncertainty. From an insurance and risk management viewpoint, it is understandable that risk is considered to be uncertainty. That's a basic actuarial concept. But the definitions quoted do not communicate entirely the nature of risk for which safety professionals give counsel. I cannot conceive of safety professionals presenting themselves solely as consultants in uncertainty reduction.

Also, the definitions of risk given in risk management and insurance literature seldom mention the severity of an event's consequences—even by implication. Giving advice to reduce the severity of results of an incident is a significant part of the work of safety professionals.

Other authors include concepts of both incident probability and severity of consequences in their definitions of risk. In *Of Acceptable Risk: Science and the Determination of Safety* (5), William W. Lowrance writes that: "Risk is a measure of the probability and severity of adverse effects." William D. Rowe, in *An Anatomy of Risk* (9), gives this definition of risk: "Risk is the potential for realization of unwanted, negative consequences of an event." Rowe supports Lowrance's definition.

Expanding on Lowrance and Rowe, this definition of risk relates more precisely to the work of safety professionals:

> Risk is defined as a measure of the probability of a hazards-related incident occurring, and the severity of harm or damage that could result.

That definition requires both a measure of the probability of an incident occurring and the severity of its adverse results. It promotes a thought process that asks: Can it happen? What is exposed to harm or damage? What is the frequency of endangerment? What will the consequences be if it does happen? How often can it happen?

Professional safety practice requires addressing the two distinct aspects of risk:

- avoiding, eliminating, or reducing the *probability* of a hazards-related incident occurring; and
- minimizing the *severity* of adverse results, if an incident occurs.

Defining Hazards

Having defined risk, these questions should then be asked: What is the source of risk? What presents the probability of incidents occurring that could result in harm or damage?

The source of risks is hazards. Hazards are the justification for the existence of the entirety of the practice of safety.

> A hazard is defined as the potential for harm: Hazards include the characteristics of things and the actions or inactions of people.

Risk was defined as a measure of the probability of a hazards-related incident occurring and the severity of harm or damage that could result. All risks with which safety professionals are involved derive from hazards. There are no exceptions.

Our Baffling and Nondescriptive Titles

Whatever safety professionals call themselves, the generic base of their existence is hazards. If there are no hazards, there is no need for safety professionals. Those statements apply, whatever words safety practitioners use in their titles—loss control, safety, risk control, environmental affairs, loss prevention, safety engineering, occupational health, et cetera.

Titles used by safety personnel may be a hindrance in achieving an understanding of the practice of safety by those outside the profession. Safety practitioners call themselves by too many names, some of which do not communicate a favorable image of what they do.

An informal and unscientific study was conducted to assess how some of the titles safety practitioners use were perceived by management personnel. Risk Managers were approached who had on their staffs people with titles like *Director of Loss Control, Director of Loss Prevention, Industrial Hygienist, Safety Manager, Director of Safety,* and *Fire Protection Engineer.*

Risk Managers arranged communications for me with their bosses or their bosses' bosses. Discussions were held to determine whether there was an understanding of what the people did who had the titles previously cited.

For the title *Fire Protection Engineer,* there was very good recognition as to function and purpose. The work of those who had the titles *Director of Safety* and *Safety Manager* was quite well understood, but not as well as *Fire Protection Engineer.*

Unfortunately, the title *Industrial Hygienist* got the least recognition and was often equated with sanitation. As a part of a title, *Occupational Health* was frequently well understood as to role and purpose.

Loss Control and *Loss Prevention* as titles did not convey clear images of purpose, and recognition of the function of those having such titles was poor. *Loss Control* was often believed to represent the security function of inventory control. On several occasions, *Loss Prevention* was assumed to be a part of claims management.

Loss Control and *Loss Prevention* as functional designations have their origins in the insurance business. Within the insurance fraternity and among some other safety practitioners, the terms are understood. But those terms do not convey clear messages of purpose and function to people outside that group.

If I had a magic wand with which I could eliminate the use of the titles *Loss Control, Loss Prevention,* and *Industrial Hygienist* by those engaged in the practice of safety, I would do so. And I would believe that I had performed a highly beneficial service. If the names safety professionals give themselves baffle people, can what they do ever be considered a profession?

Defining the Practice of Safety

Roger L. Brauer's definition of safety engineering, in *Safety and Health for Engineers* (10), is a good place to start in developing a definition of the practice of safety.

> Safety engineering is the application of engineering principles to the recognition and control of hazards.

A compatible definition appears in *Introduction to Safety Engineering* (3) by Gloss and Wardle.

> Safety engineering is the discipline that attempts to reduce the risks by eliminating or controlling the hazards.

At a meeting of the Board of Certified Safety Professionals, a definition of safety practice was written during discussions of a project to validate that examinations given properly measure what safety professionals actually do. It is as follows:

> Safety practice is the identification, evaluation, and control of hazards to prevent or mitigate harm or damage to people, property, or the environment. That practice is based on knowledge and skill as respects Applied Engineering, Applied Sciences, Applied Management, and Legal/Regulatory and Professional Affairs.

Dr. Thomas A. Selders, CSP, CIH, PE, has said that the practice of safety should be anticipatory and proactive. I support that idea. Recognizing the definitions of safety practice and of safety engineering I have cited, and Dr. Selder's comments, I propose adoption of this definition of the practice of safety.

The Practice of Safety

- serves the societal need to prevent or mitigate harm or damage to people, property, and the environment, deriving from hazards;
- is based on knowledge and skill as respects Applied Engineering, Applied Sciences, Applied Management, and Legal/Regulatory and Professional Affairs;
- is accomplished through
  - anticipating, identifying, and evaluating hazards, and
  - actions taken to avoid, eliminate, or control those hazards;
- has as its ultimate purpose attaining a state for which the risks are judged to be acceptable.

While safety professionals may undertake many tasks in their work, the underlying purpose of each task is to have the attendant risks at an acceptable level. Every element of a safety initiative should relate to hazards and the risks that derive from them.

This Definition of the Practice of Safety Applies Broadly

Whatever the particular field of endeavor and the name it is given, the entirety of the practice of safety is represented in this definition. It applies to all occupational fields for which the generic base is hazards: occupational safety, occupational health, environmental affairs, product safety, all aspects of transportation safety, safety of the public, health physics, system safety, fire protection engineering, et cetera.

To all for whom the generic base of their existence is hazards, a previously made statement applies. If there are no hazards, there is no need for their existence.

Major Elements in the Practice of Safety

There are three major elements in the practice of safety, all hazards-focused:

- in the design processes, pre-operational—where the opportunities are greatest for identifying and analyzing hazards, and for their avoidance, elimination, or control;
- in the operational mode—where, integrated within a continuous improvement process, hazards are identified, evaluated, eliminated, or controlled, *before* their potentials are realized and hazards-related incidents occur; and
- post incident—through investigation of hazards-related incidents to determine and eliminate or control their causal factors.

A Good Position Description

Mention is made in more than one place in this book of the *Scope and Functions of the Professional Safety Position* (11). It was reissued by the American Society of Safety Engineers in 1993. As a position description, it is exceptionally well done. Because of its thoroughness and accuracy, I recommend it as a knowledge source and as reference. My belief is that the definition I have given of the practice of safety is in concert with the *Scope and Functions of the Professional Safety Position.*

Hazards Management—an Umbrella Designation

I recognize that the term "the practice of safety" would probably not be easily accepted as encompassing all functions for which the generic base is hazards. Thus, "hazards management" is offered as an appropriate umbrella designation for all practices for which the generic base is hazards.

Knowledge and Skill Requirements

The knowledge and skill requirements to enter the practice of safety and to fulfill the requirements of professional safety practice are discussed in Chapter 4,

"Academic and Skill Requirements for the Practice of Safety." Those knowledge and skill requirements are, of necessity, exceptionally broad.

Conclusion

If a mission statement were written to establish the purpose of the practice of safety within an entity's goals, the following premise will serve well as a reference.

> The entirety of purpose of those responsible for safety, in fulfilling their societal responsibilities, is to manage their endeavors with respect to hazards so that the risks deriving from those hazards are at an acceptable level.

It is the intent of this essay to define the practice of safety in a logical and precise manner. All safety professionals who would like their practices to be thought of as representing a profession are invited to move this discussion forward.

References

1. Bickelhaupt, David L. *General Insurance.* Homewood, IL: Richard D. Irwin, Inc., 1983.
2. *1989 Annual Report.* National Safety Council. Itasca, IL, 1989.
3. Gloss, David S., and Miriam Gayle Wardle. *Introduction to Safety Engineering.* New York: John Wiley & Sons, 1984.
4. Hammer, Willie. *Occupational Safety Management and Engineering.* Englewood Cliffs, NJ: Prentice-Hall, 1985.
5. Lowrance, William W. *Of Acceptable Risk: Science and the Determination of Safety.* Los Altos, CA: William Kaufman, Inc., 1976.
6. Mehr, R.I., and E. Commack. *Principles of Insurance.* Homewood, IL: Richard D. Irwin Inc., 1976.
7. Greene, M. R. *Risk and Insurance.* Cincinnati: South Western Publishing Company, 1979.
8. Williams, Jr., C. Arthur, and Richard M. Heins. *Risk Management and Insurance.* New York: McGraw-Hill, 1985.
9. William D. Rowe. *An Anatomy of Risk.* New York: John Wiley & Sons, 1977.
10. Brauer, Roger L. *Safety and Health for Engineers.* New York: Van Nostrand Reinhold, 1990.
11. *Scope and Functions of the Professional Safety Position.* Des Plaines, IL: American Society of Safety Engineers, 1993.

Chapter 4

Academic and Skill Requirements for the Practice of Safety

INTRODUCTION

Having defined the practice of safety in Chapter 3, these two subjects are now addressed:

- the academic knowledge and skills that would prepare one to enter the practice of safety; and
- the knowledge and skill requirements that describe the applied practice of safety, using the term broadly.

To determine what is being taught about the knowledge and skill requirements for the practice of safety, I reviewed eighteen textbooks commonly used in safety curricula. That exercise was not productive. In fairness, the authors of the texts reviewed may not have intended to define the knowledge and skill requirements for the practice of safety.

These observations were made: Texts which focused on a specialized subject fulfilled their purposes very well. It is understandable that they emphasized the knowledge and skill requirements for that subject. Some of the texts accented safety management methods. Others related primarily to engineering practice. Four of the eighteen texts presented a good balance of safety technology and safety management.

Fortunately, the work of many others, done under the auspices of the American Society of Safety Engineers (ASSE) and the Board of Certified Safety Professionals (BCSP), serves as a valuable resource for the purposes of this essay.

Evolution of Joint ASSE and BCSP Baccalaureate Curriculum Standards

For many years, ASSE accredited college and university safety degree programs that met its standards. Over time, members of the ASSE board of directors and others contributed to the development of the "Suggested Core Curriculum for the Occupational Safety and Health Professional" that was used in the accreditation process.

In April 1980, BCSP issued Technical Report No. 1, *Curricula Development and Examination Study Guidelines* (1). This report was issued in response to a "perceived need to compile and publish data that would provide insight into the academic preparation and experience" required by those who would take what was then a single level BCSP examination. Ralph J. Vernon, Ph.D., CSP, was the author.

Dr. Vernon had reviewed the education and experience of the 4,400 applicants who had, up to that time, been examined by BCSP. He also had available two research papers that were developed for the National Institute for Occupational Safety and Health (NIOSH). One was *A Nationwide Survey of the Occupational Safety and Health Work Force* (2). The other was *Development and Validation of Career Development Guidelines by Task/Activity Analysis of Occupational Safety and Health Professions: Industrial Hygiene and Safety Profession* (3).

As a part of his work, Dr. Vernon prepared and included in the BCSP Technical Report No. 1 a "Suggested Baccalaureate Curriculum for the Safety Professional." (1) In August 1981, BCSP published an extension of the original curriculum guideline in Technical Report No. 2. Its title is *Curricula Guidelines for Baccalaureate Degree Programs in Safety* (4). That report was prepared by the BCSP Ad Hoc Committee on Academic Guidelines. Both of these reports are still valuable resources.

Distribution by BCSP of those reports had a broad impact on many safety professionals, including me. At the time of their publication, I was a member of the board of directors of ASSE. Working with other board members, modifications were proposed and made in the ASSE curriculum guidelines that moved them closer to those of the BCSP.

Curriculum Standards for Baccalaureate Degrees in Safety (5) were published in August 1991 by ASSE and BCSP as Joint Report No. 1. Subsequently, ASSE arranged with the Accreditation Board for Engineering and Technology (ABET) to have ABET's Related Accreditation Commission (RAC) be the accrediting entity for college and university safety degree programs. The ASSE/BCSP *Curriculum Standards for Baccalaureate Degrees in Safety* form the basis for accreditation by ABET/RAC.

Do ASSE/BCSP Curriculum Standards Match Educational Needs?

Just how well would graduates be prepared to enter the practice of safety if they fulfilled the course requirements of the ASSE/BCSP Curriculum Standards? In making that judgment, one would have to consider the recent transitions in the responsibilities of safety professionals, the diversity of their roles, and the many specialties within the practice of safety.

As an indication of those transitions, the responsibilities of many safety professionals have been extended and now include occupational safety, occupational health, and environmental affairs. Some are also engaged in other hazards-related functions such as fire protection, product safety, and transportation safety.

No college curriculum could include in a baccalaureate safety degree program all the courses that one would like students to take in preparation for entry into the practice of safety. So, at a baccalaureate level the course work should be basic and preparatory, and provide broad opportunities in anticipation of current and evolving professional needs.

(Assuming no practical limitations when theorizing on the ideal outline of courses to be taken by a student preparing for the practice of safety, it was easy to get the course load up to 180 hours. And that's absurd.)

I have made a thorough study of the *Curriculum Standards for Baccalaureate Degrees in Safety*. They provide an excellent foundation for the diverse needs of those entering the safety profession. Of course, modifications to reflect the changes in professional practice that occur will always be necessary. ASSE and BCSP are now making a review of the Curriculum Guidelines for updating purposes.

The Standards are foundational and their content is very broad. They recognize the need to have several electives available in specialty fields, and that individual universities will have particular course requirements.

Several outstanding safety professionals contributed to the development of the *Curriculum Standards for Baccalaureate Degrees in Safety* and they should be commended for their work. Both ASSE and BCSP should be proud of their accomplishments.

A Review of the Curriculum Standards

To meet the ASSE/BCSP accreditation requirements for a baccalaureate degree in safety, a college or university program must require not less than 120 semester hours (or equivalent) of study and meet a broad range of prescribed studies. They are briefly reviewed here.

Minimum requirements for lower-level courses are listed under the caption "University Studies." Those subjects usually exist in a variety of departments. Six foundational course categories are considered essential for the safety professional. They are listed here, along with specific course requirements.

- *Mathematics, Statistics, and Computer Science*
  Courses required: Calculus, Statistics, and Computer and Information Processing.

- *Physical, Chemical, and Life Sciences*
  Courses required: Physics (with laboratory), Chemistry (with laboratory) and Life Science (a laboratory component is recommended). An Organic Chemistry course is strongly recommended.
- *Behavioral Science, Social Science, and Humanities*
  Courses required: Psychology, and other Social Science and Humanities courses, totaling at least fifteen semester hours.
- *Management and Organizational Science*
  Course strongly recommended: Introduction to Business or Management. Course recommended: Business Law or Engineering Law.
- *Communication and Language Arts*
  Courses required: Rhetoric and Composition, and Speech. A course in technical writing is strongly recommended.
- *Basic Technology and Industrial Processes*
  Courses required: Applied Mechanics, or its equivalent, and Industrial or Manufacturing Processes.

A program would be considered deficient if it did not contain a "strongly recommended" course, unless its absence could be well supported.

Next in the Curriculum Standards is a listing of "Professional Core" courses that are to develop the basic knowledge and skills required in the practice of safety. All of the following are required courses.

- Introduction to Safety and Health
- Safety and Health Program Management
- Design of Engineering Hazard Control
- Industrial Hygiene and Technology
- Fire Protection
- Ergonomics
- Environmental Safety and Health
- System Safety and Other Analytical Methods for Safety
- Experiential Education—there shall be an internship or COOP course

"Required Professional Subjects" are listed next, and they follow. However, it is not necessary that a full course be devoted to each subject, although a school has that option.

- Measurement of Safety Performance
- Accident/Incident Investigation and Analysis
- Behavioral Aspects of Safety
- Product Safety
- Construction Safety
- Educational and Training Methods for Safety

Then, "Professional Electives" and "General Electives" follow in the Curriculum Standards. Thirty-one possible elective subjects are listed as examples, with the indication that the courses a school offers need not be limited to the listing. General electives are to fulfill the additional course hours required.

Considering the recent transitions in the practice of safety and the breadth of present and probable future requirements, I believe that the *Curriculum Standards for Baccalaureate Degrees in Safety* (5) represents a sound course of study through which the needed academic knowledge can be acquired to prepare those entering the practice of safety.

Relationship Between the ASSE/BCSP Baccalaureate Curriculum Standards and the BCSP Safety Fundamentals Examination

The Safety Fundamentals Examination (6) given by BCSP is the first of two examinations leading to the Certified Safety Professional designation. It is basic and designed to measure the broad spectrum of academic knowledge required for entry into the professional practice of safety. Only limited experience is represented in the examination.

Content of the examination should relate directly to the ASSE/BCSP Curriculum Standards if those Standards present a sound course of study. And the intent of the examination does. Members of the Board of Certified Safety Professionals who developed the Safety Fundamentals Examination did exceptional work.

Concepts on which the examination are based were established long before I served as a member of the BCSP. Although the examination is periodically validated to assure that a proper emphasis is given to each of its subjects, the principal themes remain the same. The six major sections in the Safety Fundamentals Examination and the percentage of questions in the examination for each section are: Basic and Applied Sciences (25 percent); Program Management and Evaluation (18 percent); Fire Prevention and Protection (14 percent); Equipment and Facilities (19 percent); Environmental Aspects (14 percent); and System Safety and Product Safety (10 percent). A brief review follows of its "Subject Definitions."

- *Basic and Applied Sciences*
  Basic sciences are the mathematical, natural, and behavioral sciences necessary to provide a fundamental understanding of natural and behavioral phenomena. Subjects include Mathematics, Physics, Chemistry, Biological Sciences, Behavioral Sciences, Ergonomics, Engineering and Technology, and Epidemiology.
- *Program Management and Evaluation*
  This domain encompasses the managerial, legal, philosophical, and ethical concepts and methods governing the management, administration, and

evaluation of safety programs. Subjects include Organization, Planning, and Communication; Legal and Regulatory Considerations; Program Evaluation; Disaster and Contingency Planning; and Professional Conduct and Ethics.

- *Fire Prevention and Protection*
  Fire prevention and protection encompasses the principles and methods of preventing and protecting against fires and explosions to control and contain losses to personnel and property. Subjects include Structural Design Standards, Detection and Control Systems and Procedures, and Fire Prevention.
- *Equipment and Facilities*
  The equipment and facilities section addresses exposure to and control of traumatic and cumulative injury-producing conditions. Subjects include Facilities and Equipment Design, Mechanical Hazards, Pressures, Electrical Hazards, Transportation, Materials Handling, and Illumination.
- *Environmental Aspects*
  This section addresses the recognition, evaluation, and control of environmental exposures which may result in injury or illness in the workplace, but covers public and consumer protection as well. Subjects include Toxic Materials, Environmental Hazards, Noise, Radiation, Thermal Hazards, and Control Methods.
- *System Safety and Product Safety*
  This section addresses the principles of system safety analysis as applied to systems, processes, and products and the principles of safe product design and product liability prevention. Subjects include Techniques of System Safety Analysis, Design Considerations, Product Liability, Reliability and Quality Control.

Obviously, the content of the BCSP Safety Fundamentals Examination resembles closely the requirements of the ASSE/BCSP Curriculum Standards.

Knowledge and Skill Requirements for the Applied Practice of Safety

Now, on to the knowledge and skill requirements for the applied practice of safety. An emphasis on "applied" is appropriate. The BCSP Safety Fundamentals Examination tested basic academic knowledge. Limited experience is represented in the examination. The Comprehensive Practice Examination (6) given by BCSP tests the applied knowledge and skill gained through the experience of professional safety practice. It requires a greater depth of knowledge than is required for the Safety Fundamentals Examination. All candidates for the Certified Safety Professional designation must take the Comprehensive Practice Examination.

A validation review was recently completed by BCSP: Safety professionals were asked about their work and what portion of their practices relates to each major examination subject, and about the knowledge and skill requirements for the individual practice fields within those subjects. Comments by respondents on how the requirements of their jobs had changed were considered in making revisions in examinations.

I believe that the BCSP Comprehensive Practice Examination properly represents the applied practice of safety. Fields of knowledge and skill represented in the examination are exceptionally broad. It's doubtful that a safety professional would have deep knowledge and skill in all of the examination subjects. No one gets close to a perfect score.

The four major domains and the percent of questions in the examination for each domain are: Engineering (30 percent); Management (35 percent); Applied Sciences (15 percent); and Legal/Regulatory Aspects and Professional Conduct and Affairs (20 percent).

Brief comments follow on the four domains and their knowledge and skill rubrics.

Engineering

This domain is concerned with the application of the sciences for the safe design of systems, processes, equipment, and products.

Safety Engineering involves engineering considerations related to the control of traumatic and cumulative injury-causing exposures and property damage. Knowledge fields include engineering design methods, engineering mechanics, geotechnics/soil mechanics, structural systems, electrical systems, mechanical systems, materials handling, inspection and control procedures, ergonomics and human factors engineering, facilities planning and layout, and inferential statistics.

Fire Protection Engineering is concerned with the application of fire protection engineering methods to the safeguarding of life and property against loss from fire, explosion, and related hazards. Knowledge fields include fire protection design parameters, fire detection systems, fire extinguishing systems, and process fire hazard control.

Occupational Health Engineering addresses the application of engineering methods to eliminate or control exposures to environmental agents or stresses arising in work environments that may result in impaired health. Knowledge fields include process design parameters, industrial ventilation, noise control methods, radiation protection design parameters, personal protective equipment, and chemical hazard protection.

Product and System Safety Engineering deals with the design and manufacture of systems and products to eliminate or control hazards to the user of the product or system. Knowledge fields include qualitative hazard analysis, quantitative hazard analysis, quality control and reliability, design parameters, and maintainability.

Environmental Engineering addresses the application of engineering methods to analyze and eliminate, control or remedy hazards of environmental agents that may endanger the public, air, water, soils, or the environment in general. Knowledge fields include environmental engineering design, environmental analysis methods, waste management, control and remediation, air quality management, and water quality management.

Management

This domain is concerned with the application of general management principles and techniques to the management of safety programs and the safety function.

Applied Management Fundamentals addresses the application of fundamental management principles to managing safety programs and the safety function. Knowledge fields include general principles of management, economic analysis, business law, and safety management techniques.

Business Insurance and Risk Management addresses the concepts and methods in business insurance and risk management that have direct application to safety. Knowledge fields include property and casualty insurance, workers compensation, product liability, and concepts of risk management.

Industrial and Public Relations includes elements of industrial and public relations applicable to safety management. Knowledge fields include public relations, labor relations, personnel management, and safety training.

Organizational Theory and Organizational Behavior is concerned with the design and function of various organizational structures and the behavior of individuals in organizations. Knowledge fields include organizational theory, types of organization structures, organizational behavior, and organization of the safety function.

Quantitative and Qualitative Methods for Safety Management addresses the application of quantitative and qualitative approaches to the planning, evaluation, analysis, documentation, and decision-making processes related to the management of the safety function. Knowledge fields include probability theory, descriptive and inferential statistics, computer science, mathematic methods for decision making, cost-benefit analysis, and qualitative methods.

Applied Sciences

This domain is concerned with the application of scientific principles to the recognition, assessment, and control of hazardous exposures.

Chemistry is concerned with the application of knowledge of the composition and transformations of matter for the assessment and control of hazards.

Physics addresses the non-engineering application of mechanics, heat, electricity, magnetism, wave motion, sound, and ionizing and nonionizing radiation to the assessment and control of hazardous exposures.

Life Sciences is concerned with the application of the biological sciences to protect individuals from the effects of pathogens and chemical and physical agents.

Behavioral Sciences addresses the application of psychology and sociology to the assessment of human behavioral capabilities and limitations, motivation, and behavior modification.

Legal/Regulatory Aspects and Professional Conduct and Affairs

This domain addresses safety legislation and associated standards and regulations, liability considerations, professional ethics, and professional concerns.

Legal Aspects is concerned with understanding the impact of requirements imposed by safety legislation, and the liability implications of tort and administrative law as applied to safety. Knowledge fields include legislative acts, liability, tort law, and administrative law.

Regulatory Aspects is concerned with the promulgation and administration of regulation under legislative acts dealing with safety and with the concepts of consensus standards. Knowledge fields include regulatory agencies, mandatory standards, and voluntary and consensus standards.

Professional Conduct and Affairs addresses the ethical and legal concerns of the safety professional relative to society and his/her colleagues. Knowledge fields include professional ethics, interpersonal relations, legal considerations, professional organizations, and significant developments—historical and current.

Without question, the content of the BCSP Comprehensive Practice Examination represents a great breadth of "applied" knowledge and skill. And it realistically relates to the professional practice of safety.

CONCLUSION

Every recognized profession has developed a body of knowledge and skills that is unique to that profession. It is to the advantage of safety professionals, in seeking professional recognition, to promote a course of study representing the appropriate body of knowledge for those entering the practice of safety, and to establish broadly the knowledge and skill standards for applied safety practice. For those purposes, the work done by ASSE and BCSP is commendable.

POSTSCRIPT

ASSE and BCSP have also developed and published Curriculum Standards for Master's Degrees. A March 1994 publication, Joint Report No. 2, is titled *Curriculum Standards For Master's Degrees in Safety* (7). Accreditation has been given by ABET/RAC to several Master's Degree Programs.

ASSE and BCSP issued Joint Report No. 3 in November 1994. Its title is *Curriculum Standards for Safety Engineering Master's Degrees and Safety Engineering Options in Other Engineering Master's Degrees* (8). A baccalaureate degree from an ABET accredited engineering program is one of the prerequisites for this Master's Degree. No such degree program has yet been accredited by ABET/RAC, although several schools are active candidates.

REFERENCES

1. Vernon, Ralph J. *Curriculum Development and Academic Study Guidelines.* Technical Report No. 1. Savoy, IL: Board of Certified Safety Professionals, 1980.
2. *A Nationwide Survey for the Occupational Safety and Health Work Force.* NIOSH, Publication No. 78-164, July 1978.
3. *Development and Validation of Career Development Guidelines by Task/Activity Analysis of Occupational Safety and Health Professions: Industrial Hygiene and Safety Professional.* NIOSH, Contract CDC-99-74-94, 1977.
4. *Curricula Guidelines for Baccalaureate Degree Programs in Safety.* Technical Report No. 2. Savoy, IL: Board of Certified Safety Professionals, 1981.
5. *Curriculum Standards for Baccalaureate Degrees In Safety.* Joint Report No. 1. Des Plains, IL: American Society of Safety Engineers, and Savoy, IL: Board of Certified Safety Professionals, 1991.
6. *Candidate Handbook.* Savoy, IL: Board of Certified Safety Professionals, October 1996.
7. *Curriculum Standards For Master's Degrees In Safety.* Joint Report No. 2. Des Plaines, IL: American Society of Safety Engineers, and Savoy, IL: Board of Certified Safety Professionals, 1994.
8. *Curriculum Standards For Safety Engineering Master's Degrees and Safety Engineering Options in Other Engineering Master's Degrees.* Joint Report No. 3. Des Plaines, IL: American Society of Safety Engineers, and Savoy, IL: Board of Certified Safety Professionals, 1994.

Chapter 5

On Becoming a Profession

INTRODUCTION

Safety practitioners continue to strive for recognition as a profession. They seek that recognition within society, from other professions, from their employers, and from each other. They will attain recognition as a profession only when the practice of safety meets the regimens of a profession, and only when the content and quality of their performance earn professional respect.

We who are engaged in the practice of safety use the word *professional* quite freely as a form of self-identification. For those who want to be considered safety professionals and who want the practice of safety recognized as a profession, serious introspection concerning the perceived status of what they do would serve their purposes well.

This essay will present the requirements for the practice of safety to be recognized as a profession; comments on each of those requirements; and a listing of actions that, if undertaken, would move the practice of safety toward recognition as a profession. (Insightful readers could, understandably, debate what I present.)

Obviously, we must recognize with gratitude the accomplishments over several decades by those who have been successful in promoting a higher level of preparation for, and accomplishment in, the practice of safety. Many have contributed to that progress. A challenge remains for continued gains.

A LIMITED LITERATURE REVIEW

Only a few authors have written about the practice of safety being recognized as a profession; most safety texts do not address the subject at all.

Two such articles on professional practice have appeared in *Professional Safety,* the magazine of the American Society of Safety Engineers (ASSE). In June 1981, Richard J. Finegan wrote "Is the loss control effort a profession?" (1). Dan Petersen's article "Professionalism—a fourth step" (2) was published in 1982. His article begins with this statement: "Safety is working very hard to become a profession." Petersen suggested that we should examine our theoretical base by asking whether it's fact or opinion. Many of the questions he posed in 1982 are still pertinent.

In *Techniques of Safety Management* (3), Petersen made these comments on the need for introspection:

> In the safety profession, we started with certain principles that were well explained in Heinrich's early works. We have built a profession around them, and we have succeeded in progressing tremendously with them. And yet in recent years we find that we have come almost to a standstill. Some believe that this is because the principles on which our profession is built no longer offer us a solid foundation. Others believe they remain solid but that some additions may be needed. Anyone in safety today at least ought to look at that foundation—and question it. Perhaps the principles discussed here can lead to further improvements in our approach and further reductions in our record.

In *Analyzing Safety Performance* (4), also by Dan Petersen and published in 1980, Chapter 1 is titled "The Professional Safety Task." It opens with a duplication of the then available issue of the *Scope and Functions of the Professional Safety Position* (5) which is published by ASSE.

ASSE's *Scope and Functions of the Professional Safety Position* is a widely quoted document. It has served its purpose well. It is a good job description. Those who developed the 1993 revision deserve commendation for the improvements they made.

Ted S. Ferry made this brief mention of the requirements of a profession in *Safety Program Administration for Engineers and Managers* (6).

> A profession is an occupation generally involving a relatively long and specialized preparation on the level of higher education and is usually governed by its own code of ethics. Nearly every profession has some safety and health aspects, some of them with distinct safety and health subdisciplines.

In *MORT Safety Assurance Systems* (7) by William G. Johnson, the chapter on "The Safety Function" also quotes the ASSE *Scope and Functions of the Professional Safety Position* (5).

Introduction to Safety Engineering (8) by David S. Gloss and Miriam Gayle Wardle contains the only reference I found in a safety related text that speaks of the requirements of a profession. This is what they wrote:

Hallmarks of a Profession

If safety engineering is to be considered a profession, then it must meet the criteria for professionalization. [It is] proposed that professions have specific characteristics.

1. A well-defined theoretical base
2. Recognition as a profession by the clientele
3. Community sanction for professionalization
4. A code of ethics, which regulates the professional's relationships with peers, clients, and the world at large
5. A professional organization

Dr. Roger L. Brauer, Executive Director of the Board of Certified Safety Professionals, says that you can tell that a profession has been recognized when:

- other professions seek it out;
- laws, regulations, and standards cite it; and
- there is public awareness of whom to contact for assistance to resolve a problem in the field addressed by the profession.

Requirements for the Practice of Safety to Be Recognized as a Profession, and Discussion

The previously cited references are helpful in producing the following outlines and discussions concerning the requirements for the practice of safety to be recognized as a profession.

A. Establish a Well-Defined Theoretical and Practical Base, to Include:

- a definition of the practice of safety
- the societal purpose of the practice of safety
- a recognized body of knowledge
- the methodology of the practice of safety

Discussion

"Defining the Practice of Safety," Chapter 3, was written to move the discussion forward concerning the societal purposes of the practice of safety; to establish that there is a recognized body of knowledge for the practice; to speak of the rigor of education that would prepare one to enter safety practice; and to outline its methodology.

In Chapter 10, "Applied Ergonomics: Significance and Opportunity," I refer to an article by Alphonse Chapanis titled "To Communicate the Human Factors Message, You Have to Know What the Message Is and How to Communicate It" (9). One of his themes is that human factors engineering has to be defined and its practitioners must know what it is to be able to communicate about it successfully. Safety professionals have the same need.

The practice of safety, as defined, includes all fields of endeavor for which the generic base is hazards—occupational safety, occupational health, environmental affairs, product safety, transportation safety, safety of the public, health physics, system safety, fire protection engineering, et cetera. The definition I developed of the practice of safety is also being given here as a reference for the discussion that follows.

The Practice of Safety

- serves the societal need to prevent or mitigate harm or damage to people, property, and the environment, deriving from hazards;
- is based on knowledge and skill as respects Applied Engineering, Applied Sciences, Applied Management, and Legal/Regulatory and Professional Affairs;
- is accomplished through
  - anticipating, identifying, and evaluating hazards, and
  - actions taken to avoid, eliminate, or control those hazards;
- has as its ultimate purpose attaining a state for which the risks are judged to be acceptable.

In ASSE's *Scope and Functions of the Professional Safety Position* (5), these comments are made about the education, training, and experience needs of safety professionals.

> To perform their professional functions, safety professionals must have education, training, and experience in a common body of knowledge. Safety professionals need to have a fundamental knowledge of physics, chemistry, biology, physiology, statistics, mathematics, computer science, engineering mechanics, industrial processes, business, communications, and psychology.

Note the term "a common body of knowledge." In the process of researching and writing a definition of the practice of safety, I concluded that the *Curriculum Standards For Baccalaureate Degrees in Safety* (10), jointly published by the Academic Accreditation Council of the American Society of Safety Engineers and the Board of Certified Safety Professionals, establishes a rigor of education that prepares one to enter the safety profession.

Further, it was observed that the domains and rubrics of the Comprehensive Practice Examination (11) given by the Board of Certified Safety Professionals defines the breadth of knowledge and skill required in applied safety practice.

Serious questions about our "recognized body of knowledge" continue to arise, since some safety practitioners still hold dearly to far too many myths. In *Safety Program Administration For Engineers and Managers* (6), Ted S. Ferry devotes a chapter to "Those Cherished and Hallowed Old Safety Cliches and Truisms."

We should take a professional approach to examining those safety cliches and, where appropriate, educate ourselves concerning them. Comments will be made here about only a few of those myths. Others can surely add to this list.

1. At ASSE Professional Development Conferences, one or more speakers always informs attendees that 90 percent of accidents are caused by unsafe acts of employees. How pitifully unprofessional! Heinrich's 88-10-2 theory was held as the conventional wisdom years ago. It is a shallow myth.

2. Many still offer as truth Heinrich's foundation of a major injury: the 1-29-300 premise, which states that: "in a unit group of 330 similar accidents occurring to the same person, 300 will result in no injury, 29 will produce minor injuries, and 1 will cause a serious injury." Think about that—330 similar accidents occurring to the same person. Would that include a fall off a 50-story building?

   Bird and Loftus (12) propose a different ratio: "1 disabling injury for every 100 minor injuries and 500 property damage incidents."

   Use of these statistical bases gives support to the principle that if we give adequate attention to the frequency of incidents, we will also be taking care of severity potential. That may or may not be so, depending on whether the severity potential is also represented in the more frequent incidents.

   It has not been possible to locate a body of research that supports the validity of either the Heinrich or the Bird and Loftus postulations. They are mythical. Yet safety professionals continue to offer them as truths.

3. Does the well-used axiom—garbage in, garbage out—apply to our incident analysis systems? Do we mythically hold to information-gathering systems that produce misinformation?

4. Is the occupational health exposure really as significant as many profess? For well over forty years, we have been hearing about the latency period after which occupational illnesses would appear in great numbers as workers compensation claims. That has not happened. Yes, there are many real occupational health exposures. But, has the time come to wonder about their actual extent? Some scientists now question previous decisions and the validity of models used in arriving at health risks.

   In an exhibit titled "Nonfatal Occupational Injuries Involving Days Away from Work" in the National Safety Council's 1996 issue of *Accident*

> *Facts* (13), "Exposure to harmful substance" as an event or exposure category represented 5 percent of total injuries and illnesses. My first study of the magnitude of the occupational health exposure was made over twenty years ago. I concluded then that occupational illness cases represented 5 percent of the total. That percentage has not changed over the years.

Has the time arrived for safety practitioners to stop reciting cliches, repeating the literature—without requiring substantiation? Should we cease docilely adopting published premises without promoting scientific inquiry as to their validity?

In no way is it intended that the preceding comments be considered all inclusive, especially coming from just one source.

B. Developing a Common Language within Safety Practice, with a Realization That:

- a "hallmark" of a profession is to define itself;
- the public does not know who we are or what we do; and
- we confuse others with our multitude of titles.

Discussion

Try this experiment. Have your associates assume that they are asked by a member of the public: what is your job? What do you do? I asked those questions of a group of safety professionals, and the answers were embarrassing. We have not established a common understanding of our practice, nor do we use a common language to define what we do. If we are to be recognized as a profession, we must be able to identify what is unique about it, and its societal purposes.

One of the objectives of Chapter 3, "Defining the Practice of Safety," is to emphasize the significance of our not having yet defined our practice. And, we must do so to attain professional recognition.

Names of other professions—law, medicine—immediately bring to mind a mental image of that field of endeavor, and the requirements to be a member of the profession. We must follow that lead and define who we are. What we call ourselves and the language we use in communications both with our clients and in the community at large should convey an appropriate image of our discipline. The great variations in terms and titles we use may confuse those with whom we try to communicate. There is no question that we baffle decision makers, and the public, with the multitude of titles we give ourselves.

A brief and unscientific study was made of the understanding decision makers have of the titles we use (see Chapter 3, "Defining the Practice of Safety"). I wrote that if I had a magic wand with which I could eliminate the use of the titles *Loss Prevention, Loss Control,* and *Industrial Hygiene,* I would do so and believe that I had done a great service for those engaged in the practice of safety.

In addition, I said that the generic base of the practice of safety is hazards, and that if there were no hazards, there would be no reason for safety professionals to exist. In time, we will more than likely be considering an umbrella term that encompasses all aspects of safety practice. Hazards Management would serve very well for that purpose.

C. Achieving Recognition as a Profession by the Clientele to Whom We Give Advice, Considering:

- the content, the substance, of the advice we give;
- whether the advice we give achieves expected results; and
- the nature of our communication.

Discussion

It's possible that the substance of the communications of too many safety professionals to decision makers is perceived as shallow, superficial, and not pertinent to an entity's real hazards management needs. Much of our language developed years ago. It is time to evaluate the real substance of it. As an example, are the terms *unsafe act* and *unsafe condition* obsolete?

Too many safety professionals are perceived as having established a culture of separateness, and of having purposes that do not directly help management attain their goals. Safety professionals should strive to be perceived as part of the management team, and as cognizant of the goals of, and the constraints on, the organizations for which they provide a consultancy.

Our literature frequently indicates that the causes for all incidents resulting in injury or illness derive from *something being wrong in the management system.* And we promote the idea, over and over. As we do so, are we being overly critical of a group of which we want to be a part? Do we gain more or lose more by its never-ending repetition?

Yes, it can be theorized that there was a management shortcoming for every hazard, the realization of which resulted in injury or damage. And it's necessary to identify those shortcomings in causal studies. But acceptance by our clientele can be obtained more effectively through language that demonstrates participation toward achieving goals we share.

At our professional conferences, we often discuss "how to obtain management support." Speakers address effective communication methods, and they are important. But I suggest that we will gain from an examination of the content of our practice and the substance of our communications. If a practice is not perceived as being professional and having import, the world's best communicator will still not get the management support and involvement necessary.

D. Promoting and Supporting Research, Recognizing That:

- knowledge requirements concerning hazards and the methodologies to anticipate, avoid, eliminate, or control those hazards will continue to expand; and
- it is typical of a profession to continuously examine the effectiveness and outcomes of the solutions proposed, and seek to innovate.

Discussion

How much should be said about the expanded knowledge needs of safety professionals, especially in the past ten years? It should concern us that safety research is most often done by people who would not consider themselves to be safety professionals. We are almost entirely excluded from determining what the research needs are, and from assessing the results.

This is a subject for which, with rare exceptions, activity by safety practitioners will be an original undertaking. As a beginning, research is needed to evaluate the premises that have accumulated in the practice of safety in the past seventy-five years. If we are to be recognized as a profession, we must establish that we are promoters and supporters of research.

E. Maintaining Rigid Certification Requirements, Promoting the Significance of Certification, and Giving Additional Status to Certification.

Discussion

Much thanks should be given to those visionaries on the Board of Directors of the American Society of Safety Engineers in the 1960s who conceived of and established the Board of Certified Safety Professionals (BCSP). As a result of their work, a sound and proven certification program exists for safety professionals, for which recognition continues to grow. The CSP designation (the Certified Safety Professional designation) has become the mark of the professional within the safety field.

Little will be said here about the BCSP program. Initiated in 1969, it has stood the test of time. The number of those seeking examinations and certification has steadily increased. In its determination to be current, BCSP undertook a validation project to assure that its examinations properly examine the practice of safety as it has evolved. That resulted in examination rubrics being expanded.

All involved—individual safety professionals, the BCSP, ASSE—could do a better job of promoting the significance of certification and of giving it a higher status. Individuals who employ safety professionals can especially give greater significance to the CSP designation.

F. Adhering to an Accepted Standard of Conduct, Which Is an Absolute Requirement of a Socially Recognized Profession.

Discussion

How can anyone presume to be a professional without being willing to meet high standards of performance and to insist that others in the profession do the same? A prescribed statement of professional conduct would cover the relationships expected in individual practice with one's clients, peers, and the community at large.

The American Society of Safety Engineers and the Board of Certified Safety Professionals have issued Codes of Professional Conduct which deserve publicity and promotion.

G. Having a Professional Society, Participating in It, and Supporting It.

Discussion

Several related societies exist that safety professionals can well support. Low levels of participation in such societies lead to the observation that more safety practitioners call themselves safety professionals than should. Just being a member of a professional society doesn't really qualify one for professional status.

Those who seek professional status and would like their practices to be recognized as a profession must be more prominent participants in the societies of which they are members.

H. Obtaining Societal Sanction for Professionalization.

Discussion

This is a future goal. It will have been achieved when the public perceives that the practice of those who designate themselves as safety professionals has a distinct value to society. Safety professionals will earn that respect and recognition only through their performance.

When that occurs it will be expected that a person with a prescribed professional education, experience, and certification will fulfill safety responsibilities. We've come a long way and have a long way to go, especially with respect to recognition by the general public.

At least in employment, the value of the CSP designation has grown, which is a form of societal sanction. There has been a continual increase in the percentage of job advertisements for safety positions that mention the desirability of being a Certified Safety Professional.

Conclusion

This essay is intended to promote individual and collective introspection. Accomplishment for each of the action subjects discussed in this essay will serve to achieve professional status for those engaged in the practice of safety, which is an ultimate ideal. A summary pertaining to those action subjects follows.

Action Subjects

1. Agreeing on and promoting an understanding of a definition of the practice of safety and its basic methodology.
2. Determining that the *Curriculum Standards for Baccalaureate Degrees in Safety* jointly published by ASSE and BCSP represent sound preparation for one to enter the safety profession, and strongly promoting their extended adoption.
3. Establishing that the domains and rubrics of the BCSP Comprehensive Practice Examination represent the breadth of knowledge and skill required for safety practice, and communicating that to the public and to safety practitioners.
4. Examining safety literature to identify what is patently unprofessional, with the intent of arranging exposition and debate on those subjects.
5. Promoting adoption by safety professionals of methods of scientific inquiry, and encouraging peer review and verification of that which is published.
6. Establishing a system to review the knowledge fields for which additional information is needed to maintain professional practice, and arranging development and dissemination of that information.
7. Undertaking to develop a commonly accepted language that clearly presents an image of the practice of safety.
8. Arranging for a study of the public understanding of the titles we use, and promoting the use of those which best convey the image of a profession.
9. Promoting exploration of incident causation theory, current knowledge of which is both fundamental and vitally needed for the practice of safety.
10. Taking the initiative in arranging, promoting, and supporting research projects.
11. Promoting the significance of the CSP designation by individual safety professionals, BCSP, and ASSE.
12. Publicizing the ASSE and BCSP standards of professional conduct, and encouraging safety professionals to consider them as foundational in their own practices.
13. Convincing safety professionals who want recognition as a professional that they should be active participants in their professional societies.

References

1. Finegan, Richard J. "Is the Loss Control Effort a Profession?" *Professional Safety* (June 1981).
2. Petersen, Dan. "Professionalism—A Fourth Step." *Professional Safety,* (November 1982).
3. Petersen, Dan. *Techniques of Safety Management.* Goshen, NY: Aloray, 1989.
4. Petersen, Dan. *Analyzing Safety Performance.* New York: Garland STPM Press, 1980.
5. *Scope and Functions of the Professional Safety Position.* Des Plaines, IL: American Society of Safety Engineers, 1993.
6. Ferry, Ted S. *Safety Program Administration for Engineers and Managers.* Springfield, IL: Charles C. Thomas, 1984.
7. Johnson, William G. *MORT Safety Assurance Systems.* New York: Marcel Dekker, 1980.
8. Gloss, David S., and Miriam Gayle Wardle. *Introduction to Safety Engineering.* New York: John Wiley & Sons, 1984.
9. Chapanis, Alphonse. "To Communicate the Human Factors Message, You Have to Know What the Message Is and How to Communicate It," *Human Factors Society Bulletin* (November 1991).
10. *Curriculum Standards For Baccalaureate Degrees in Safety.* Joint Report No. 1. Des Plaines, IL: Accreditation Council of the American Society of Safety Engineers, and Savoy, IL: Board of Certified Safety Professionals, August 1991.
11. *Candidate Handbook.* Savoy, IL: Board of Certified Safety Professionals, 1996.
12. Bird, Frank E., Jr. and Robert G. Loftus. *Loss Control Management.* Loganville, GA: Institute Press, 1976.
13. *Accident Facts.* Itasca, IL: National Safety Council, 1996.

Chapter 6

Observations on Causation Models for Hazards-Related Incidents

INTRODUCTION

A safety professional who gives advice on avoiding, eliminating, or controlling hazards in any of the three elements in the practice of safety (pre-operational, operational, and post-incident) must understand how hazards-related incidents occur to be effective. It is basic in problem-solving to define and understand the problem, to analyze the cause-and-effect relationships of the subsets of the problem, to consider alternate solutions, to choose and apply the solutions, and to subsequently evaluate their efficacy.

This essay addresses the need for safety professionals to have an understanding of the hazards-related incident phenomenon and to be able to identify and analyze the cause-and-effect relationships in the subsets of the incident process.

PROFESSIONAL SAFETY PRACTICE REQUIRES ESTABLISHING AN ACCEPTED CAUSATION MODEL

Dr. Roger L. Brauer, Executive Director of the Board of Certified Safety Professionals, made this comment about proving the validity of causation models.

> Good science requires that safety professionals do the validation work necessary to prove that what they propose, based on the causation models they have adopted, is effective and that real risk reduction is achieved.

Safety professionals apply differing and contradictory incident causation models, and the work of some of them is misdirected and ineffective. Professional safety practice requires that the advice given to avoid, eliminate, or control hazards be based on a sound incident-causation thought process so that, through the application of that advice, the desired risk reduction is achieved. That will not occur if the causation model used does not require identifying the actual causal factors.

At the "Safety Technology 2000" symposium held by the American Society of Safety Engineers in June 1995, many of the papers presented made specific reference to or alluded to an accident causation concept. It was obvious, from a review of those papers, that the beliefs of safety professionals about concepts of hazards-related incident causation are far from consensus. These are the extremes of thinking on incident causation in those papers:

- 90 percent of accidents are caused by unsafe acts, and the proper solution for them is to modify employee behavior;
- causal factors for 90 percent of accidents are systemic and the proper solution for them is to modify the work system.

Assume that a given hazards-related incident is to be investigated. Safety professionals who have adopted, and give prominence and near exclusivity to, one or the other of those concepts would give greatly divergent remediation advice. The advice deriving from a narrow application of either approach would not address all of the causal factors, nor would the remedial actions proposed achieve the needed risk reduction.

If we who call ourselves safety professionals are to be truly perceived as professionals, we must resolve this matter of a generally accepted hazards-related incident causation model. A major study on this subject would be to our advantage.

Safety professionals investigating a given hazards-related incident should identify the same causal factors, allowing for an occasional exception. That is unlikely if their understandings of incident causation, and the thought processes they apply, have different and sometimes contradictory foundations.

All Incident Causation Models Can't Be Right

Ludwig Benner, Jr. wrote this in the conclusions to a study titled "Rating Accident Models and Investigation Methodologies" (1), which was undertaken for OSHA:

> The number of conceptual accident models that drive government accident investigation programs seems unnecessarily diverse. Since they conflict, all models can not be valid.

In that study, Benner rated fourteen different accident models, and seventeen accident investigation methods. The two accident models given the highest ratings were the "events process model" and the "energy flow process model"; the two accident investigation methods given the highest ratings were "events analysis" and the "MORT system." Those models and methods are related and are excellent references in developing an understanding of incident causation.

Advice given by many safety practitioners is based on the diverse and conflicting models studied by Benner—*all of which cannot be valid.*

Incident Causation Models Represent Great Diversity of Thinking

Benner rated only the models and investigation systems used in seventeen selected government agencies. At least twenty-five causation models are referenced in safety literature. They present a great diversity of thinking. These are but a few of them: single event theory; chain of events theory; epidemiological models; systems theory models; multilinear events sequencing; human factors models; life change unit theory; motivation-reward satisfaction models; and the Management Oversight and Risk Tree Model.

This quotation from a letter received from Benner gives one indication of the divergence of thinking about accident and causation models.

> . . . accident models and accident causation models involve two different areas of endeavor. The point is subtle, but in my view it is absolutely imperative to recognize the difference. Causation models purport to present cause and effects without identifying the phenomenon; no beginning and end of the phenomenon is indicated. Accident models on the other hand, deal descriptively with accidents as a process that has a beginning and an end, and the elements of that process. Please help me keep my models in the latter arena when quoting any of my work to ensure that it is not thrown into the causation model arena inadvertently.

Respectfully, an attempt will be made to comply with Benner's request. Several references will be made to Benner's work in this essay since it is considered to be important—in thinking about accident models or accident causation models.

Some Definitions

As used in this treatise, causation means the act or agency of causing or producing an effect. Causal factors include all of the elements—the events, the characteristics of things, and the actions or inactions of persons—that contribute to the incident process. A model is to represent the theoretical ideal for the process through which hazards-related incidents occur, a process that requires determining when the phenomenon begins and ends.

Recognition of the Need for an Accepted Causation Model

Several authors have recognized, with some frustration, the absence of, and a serious need for, a generally accepted accident causation model. Robert E. McClay addressed the subject in his paper titled "Toward a More Universal Model of Loss Incident Causation." (2)

> The most obvious example of a weakness in the theoretical underpinning of Safety Science is the lack of a satisfactory explanation for accident causation. . . . Line managers in an organization can be forgiven for being cynical about safety when the reasons for the occurrence of accidents seem so obscure. . . . What is needed is an acceptable model that explains the occurrence of accidental losses of all types across the entire discipline of Safety Science.

In *MORT Safety Assurance Systems* (3) by William G. Johnson, this appears:

> Improved models of the accident sequence would be helpful in understanding the dynamics of accidents and would be a basis for data collection. No fully satisfactory model has yet been developed, but many are promising and useful.

Ted S. Ferry writes this, in *Modern Accident Investigation and Analysis* (4):

> The scientific literature on mishap analysis offers little insight into the process by which mishaps occur.

Variations in current practice and the need for an accepted investigation methodology were mentioned by Benner in "Accident Investigations: Multilinear Events Sequencing Methods." (5) Benner writes:

> Approaches to accident investigations seem as diverse as the investigators. . . . The absence of a common approach and differences in the investigative and analytical methods used have resulted in serious difficulties in the safety field . . . including . . . barriers to a common understanding of the phenomenon . . . [and] popular misconceptions about the nature of the accident

> phenomenon. . . . The purpose of this paper is to call attention to the need to develop generally acceptable approaches and analysis methods that will result in complete, reproducible, conceptually consistent, and easily communicated explanations of accidents.

Defining the Incident Phenomenon

One of the difficulties to be overcome in establishing a causation model is determining what is to be encompassed and what terms are to be used in the description. This is from McClay (2).

> . . . an ideal model should be applicable across the full spectrum of Safety Science . . . it becomes necessary to use a broader term than "accident causation". . . . The term loss incident will be used . . . to include any event resulting from uncontrolled hazards, capable of producing adverse, immediate or long term effects in the form of injury, illness, disability, death, property damage or the like.

Safety professionals give many names to the incidents to which a causation model would apply: accidents, incidents, mishaps, near-misses, occurrences, events, illnesses, fires, explosions, windstorms, drownings, electrocutions, et cetera. Pat Clemens, a prominent safety consultant, has said that the language used by safety practitioners lacks words to convey precise and understood meanings. It's probable that the people with whom safety professionals try to communicate are baffled by the many terms used to describe hazards-related incidents.

As an example, it's common in safety literature to designate as "incidents" or "near-misses" the events that do not result in harm or damage, and to designate as "accidents" those that do. Differentiation is based entirely on outcomes.

Severity potential for harm or damage should determine whether a hazards-related incident is to be given priority consideration. Categorizing incidents that result in harm or damage separately from those that do not, even though the results of the latter could have been severe under slightly different circumstances, too often results in misapplication of resources.

In the term *hazards-related* it must be understood that hazards are to include any aspect of technology or activity that produces risk (Fischhoff, 6). A hazard is defined as the potential for harm: Hazards include the characteristics of things and the actions or inactions of persons.

If a hazard is not avoided, eliminated, or controlled, the potential will be realized. Whatever names we use to identify those realizations—all of the types of events previously mentioned (incidents, near-misses, accidents, explosions, electrocutions, et cetera)—they are all hazards-related incidents.

So, it is proposed that a new name be created to encompass all hazards-related incidents—HAZRIN. The term *HAZRIN* encompasses all incidents which are the realization of the potential for harm or damage, whether harm or damage resulted or could have resulted, for all fields of endeavor that are hazards related.

The Case to Be Made Here

To move the discussion forward, a case will be developed here in support of these premises:

- A practice of safety based principally on the causation model represented by the domino sequence developed by H. W. Heinrich, and the many extensions of it, that focuses on the so-called unsafe act or human error as the principal causal factor will be ineffective in relation to the actuality of causal factors.
- W. Edwards Deming got it right in *Out of the Crisis* (7) when he proposed that a very large majority of the problems in any operation are systemic, that they derive from the workplace and work methods created by management and can only be resolved by management, and that responsibility for only the relatively small remainder lies with the worker.
- Extrapolating from Deming's premise, a very large majority of the causal factors for hazards-related incidents will be systemic, and a small minority will be employee-focused.
- For almost all hazards-related incidents, even those that seem to be the least complex, there will be multiple causal factors, deriving from *less than adequate* policies, standards, or procedures that impact on workplace and work methods design, operations management, and task performance practices.
- A sound causation model for hazards-related incidents must identify and stress the significance of the design management aspects *and* the operations management aspects *and* the task performance aspects of the causal factors; and that those aspects are interdependent and mutually inclusive.

Concerning Causation Models Based on Heinrichian Principles

It is evident in recent safety literature that the prevalent causation concept applied by safety professionals derives from the writings of H. W. Heinrich. His causation concept is represented by a "domino sequence." (I still have my set of dominoes, which is forty-five years old.) A third edition of *Industrial Accident Prevention* (8) by H. W. Heinrich, published in 1950, is the source of the following:

> In the middle 1920's, a series of theorems were developed which are defined and explained in the following chapter and illustrated by the "domino sequence." These theorems show that:
>
> (1) industrial injuries result only from accidents,
> (2) accidents are invariably caused by the unsafe acts of persons or by exposure to unsafe mechanical conditions,
> (3) unsafe actions and conditions are caused by faults of persons, and
> (4) faults of persons are created by environment or acquired by inheritance.
>
> From this sequence of steps in the occurrence of accidental injury it is apparent that man failure is the heart of the problem. Equally apparent is the conclusion that methods of control must be directed toward man failure.

After presenting a graphic display of the domino sequence, this description is given of what each is to represent:

> The several factors in the accident occurrence series are given in chronological order in the following listing:
> 1. Ancestry and environment
> 2. Fault of person
> 3. Unsafe act and/or mechanical or physical condition
> 4. Accident
> 5. Injury

This is another Heinrich premise, one that is still often cited by safety practitioners:

> . . . a total of 88 per cent of all industrial accidents . . . are caused primarily by the unsafe acts of persons.

All of these excerpts from Heinrich's text focus on the individual who is involved in an incident that resulted in injury. The proposal is that in 88 percent of industrial accidents an unsafe act is committed by an employee, who has faults that derive from his or her ancestry and environment.

For years many safety practitioners based their work on Heinrich's theorems, working very hard to overcome "man failure," believing with great certainty that 88 percent of accidents were primarily caused by unsafe acts of employees. How sad that we were so much in error.

Heinrich's premises, and the several causation models that are based on them, are still the foundation of the work of many safety practitioners. Indeed, most causation models have focused on the behavior of the individual who is presumed to have acted unsafely. And many safety practitioners, in the prevention measures they propose, emphasize training, quality of leadership by supervisory personnel, behavior modification, appropriate methods of discipline—a great range of activities directed toward the control of "man failure." These solutions are to achieve a change in the performance of the employee who, when judged retrospectively, is deemed to have acted unsafely.

Use of Heinrich's ideas has led to oversimplification, and encouraged identifying a single causal factor for incidents focusing on employee error. Johnson expresses concern in this excerpt from *The Management Oversight and Risk Tree—MORT* (9) that the domino sequence might give the impression that incident causation is simplistic:

> Safety texts have used a row of falling dominoes as a model illustrating the accident sequence. This may be such a gross simplification as to limit understanding.

Unsafe Act and Unsafe Condition Are Inappropriate Terms

In the same text, Johnson questions the appropriateness of the use of the terms *unsafe conditions and unsafe acts:*

> These categories of errors are extensively used in accident data collection. However, the fact remains that concepts of unsafe conditions and unsafe acts are usually simplistic and definitions are variable from place to place.

It is not good science to use terms that cannot be defined. Definitions of unsafe conditions and unsafe acts that can withstand thorough inquiry are scarce.

I believe that the terms unsafe act and unsafe condition should be eliminated from the vocabulary of safety professionals, to be replaced by the term causal factors—which can be defined. Safety professionals should also get rid of their dominoes: they are overly simplistic representations of incident causation.

Dr. Franklin E. Mirer, Director of the Health and Safety Department of the United Auto Workers, also questions using the term unsafe acts as a causal factor (10):

> Unfortunately, the debate about whether "unsafe acts" or "unsafe conditions" cause the majority of injuries persists. No evidence supports claims that "unsafe acts" predominate as causes of injuries.
>
> . . . many of the models of causation of injuries and illnesses . . . are operator models which are sophisticated versions of "blame the worker" safety approaches.
>
> The best opportunities for prevention are at the production system level, and may be called "Design in Safety."

Often, the wrong advice is given when the causation model on which the practice of safety is based focuses primarily and almost exclusively on:

- the characteristics of the individual;
- unsafe acts being the primary causes of workplace incidents;
- corrective measures being directed principally toward effecting the behavior of the individual, to correct "man failure."

Moving the Causation Emphasis to Systems, Procedures, and the Work Environment

The definition of an accident in earlier literature on *The Management Oversight and Risk Tree—MORT* indicated that an injury was preceded by sequences of planning and operational errors which (a) failed to adjust to changes in physical or human factors, and (b) produced unsafe conditions and/or unsafe acts.

But there has been a significant and appropriate change in the MORT literature concerning the identification of causal factors. In a November 1994 publication from the Department of Energy, *Guide to Use of the Management Oversight and Risk Tree* (11), this appears under "Performance Errors":

> It should be pointed out that the kinds of questions raised by MORT are directed at systemic and procedural problems. The experience, to date, shows there are few "unsafe acts" in the sense of blameful *work level employee* failures. Assignment of "unsafe act" responsibility to a work level employee should not be made unless or until the preventive steps of: (1) hazard analysis, (2) management or supervisory detection, and (3) procedures safety reviews have been shown to be adequate.

This concept shifts the emphasis away from the employee to systems and procedures, which only management can correct.

In *Techniques of Safety Management* (12), Dan Petersen expressed a view, in his seventh of The Ten Basic Principles of Safety, concerning individual behavior being a derivative of the work environment created by management:

> In most cases, unsafe behavior is normal behavior; it is the result of normal people reacting to their environment. Management's job is to change the environment that leads to the unsafe behavior.

Petersen expanded on this thought later in his text and wrote, with significance, that:

> . . . unsafe behavior is the result of the environment that has been constructed by management. In that environment, it is completely logical and normal to act unsafely.

In the context given, just what is unsafe behavior, if it is logical and normal? Very few authors who use the term *unsafe act* define it, for a very good reason. It seems nearly impossible to arrive at a definition that is universally acceptable. And no attempt will be made to do so here.

If the environment that has been constructed, which would derive from the design of the physical aspects of the workplace, or the design of the work methods, or operational and management influences, results in logical and normal behavior that is considered unsafe, it would seem that a HAZRIN model should give prominent emphasis to the decision-making out of which those environmental causal factors arose.

Doing so would move the emphasis in the practice of safety to decisions deriving from the organization's culture, standards, and expectations that impact on systems, procedures, and the work environment, and away from outcomes such as unsafe acts.

On Work that Is Error-Provocative

Alphonse Chapanis is exceptionally well-known in ergonomics and human factors engineering circles and his writings on avoiding the design of work that is error-provocative are often cited. These are excerpts from "The Error-Provocative Situation," a chapter in *The Measurement of Safety Performance* (13).

> Design characteristics that increase the probability of error include a job, situation, or system which:
>
> a. Violates operator expectations;
> b. Requires performance beyond what an operator can deliver;
> c. Induces fatigue;
> d. Provides inadequate facilities or information for the operator;
> e. Is unnecessarily difficult or unpleasant; or
> f. Is unnecessarily dangerous.

It is illogical to conclude after an incident occurred that the principal causal factor was the unsafe act of a worker if the design of the workplace or the work methods is error-provocative. Systemic causal factors should be considered primary if the design of the work was error-provocative, or overly stressful, or if the immediate work situation encouraged riskier actions than the prescribed work methods. To identify the causal factors in such situations solely or emphatically as an "employee error," or as an "unsafe act" would be wrong and ineffective.

This is a serious subject. For incident causal factors that are actions or inactions of individuals, their so-called errors may be "programmed" into the work system created for them. And a causation model has to address the "programming" sources.

Trevor Kletz, in his book *An Engineer's View of Human Error* (14), makes this statement:

> Almost any accident can be said to be due to human error and the use of the phrase discourages constructive thinking about the action needed to prevent it happening again; it is too easy to tell someone to be more careful.

In the introduction to his book, Kletz focuses on and gives emphasis to engineers developing an understanding that humans may err and that the design of the workplace and work methods should contemplate error possibility:

> The theme of this book is that it is difficult for engineers to change human nature and, therefore, instead of trying to persuade people not to make mistakes, we should accept people as we find them and try to remove opportunities for error by changing the work situation, that is, the plant or equipment design or the method of working. Alternatively, we can mitigate the consequences of error or provide opportunities for recovery.
>
> A second objective of the book is to remind engineers of some of the quirks of human nature so that they can better allow for them in design.

Since it may occur that the design is less than adequate, does not consider people as they are, or does not anticipate possible quirky behavior, then design aspects have to be an important causal factor in a causation model. In his book, Kletz reviewed several accidents that "at first sight were due to human error" and discussed how they could have been prevented through improved design, construction, maintenance, and better management.

Kletz also suggests that we should do away with the term *human error,* since it gets in the way of inquiry to determine real causal factors.

Focusing on individual error won't necessarily lead to problem identification. Mark Paradies gives this view in his article "Root Cause Analysis and Human Factors" (15):

> . . . using the word error (as in operator error) tends to focus attention on the individual involved rather on the problem. This was driven home one day when an investigator said, "I can't list this as operator error—the operator wasn't at fault, the procedure was written wrong!"

Note—the procedure was written wrong! Focusing on improving the procedure is more effective than concentrating on individual behavior.

Giving Design Causal Factors Their Proper Place

I have written that the greatest strides forward with respect to safety, health, and the environment will be made through the design and engineering processes. In that context, processes include the design and engineering of facilities, hardware, equipment, tooling, processes, layout, work stations and the environment, and, very importantly, the design of the operating methods, and the procedures for accomplishing the work.

Most causation models that have influenced the work of safety practitioners have stressed the management and behavioral aspects of safety. A very large part of the work of safety practitioners has been directed toward improving managerial and personnel behavior. Thus, the design and engineering implications have been somewhat ignored. Safety professionals have, largely, not been involved in design and engineering, where great opportunities for achievement exist.

This essay pleads for a balanced approach that gives a proper emphasis within the practice of safety to causal factors deriving from design management, operations management, and task performance.

Because of their involvement with applied ergonomics and quality management, both of which are design-based, some safety professionals have taken a different view of causal factors. As an example, it has become apparent that for many ergonomics-related injuries and illnesses, for which the causes were classified as unsafe acts of employees, the actual causal factors were matters of workplace and work methods design.

Ergonomics problems cannot be solved, nor can superior levels of quality be achieved, without a balanced approach that includes both design and engineering operations management and task performance considerations.

David M. DeJoy alluded to the inadequacy of prevention efforts centered on "predispositions, motivations, and attitudes of workers" in "Toward a Comprehensive Human Factors Model of Workplace Accident Causation" (16).

> The essence of the contribution of human factors to safety is that machines, equipment, jobs, processes, and environments can be safer if they are designed with the capabilities and limitations of the worker in mind. While the predispositions, motivations, and attitudes of workers are important to safety performance, a comprehensive human factors analysis goes well beyond these considerations.

Key phrases in the foregoing paragraph are "designed with the capabilities and limitations of the worker in mind" and "a comprehensive human factors analysis." They suggest an anticipatory approach in the design phase and applying human factors concepts, going beyond "predispositions, motivations, and attitudes," which are basic in the Heinrichian theorems and in the causation models which are extensions of them.

Recommended Readings

It's proposed that safety professionals would benefit from an examination of the substance of the causation models on which their practices are based. For those who would undertake such an exercise toward the development of an acceptable causation model for hazards-related incidents, the following are recommended as minimal readings:

- *Investigating Accidents with STEP* (17) by Kingsley Hendrick and Ludwig Benner, Jr.
- Other papers by Benner (1, 5, 18)
- *Modern Accident Investigation and Analysis* (4) by Ted S. Ferry
- *MORT Safety Assurance Systems* (3) by William G. Johnson
- "Toward a More Universal Model of Loss Incident Causation" (2) by Robert E. McClay
- "Toward a Comprehensive Human Factors Model of Workplace Accident Causation" (16) by David M. DeJoy
- *Guide to Use of the Management Oversight and Risk Tree* (11)

There are significant commonalities and differences in these publications. My intent is to select from them to support a logical thought process, and add my own views. To begin with, a hazards-related incident, a HAZRIN, should have a definition.

Defining a Hazards-Related Incident—a HAZRIN

This definition appears in the *Guide to Use of the Management Oversight and Risk Tree* (11): "An accident is defined as unwanted transfer of energy or environmental condition because of lack or inadequate barriers and/or controls, producing injury to persons and/or damage to property or the process."

In *Investigating Accidents with STEP* (17): "An accident is a special class of process by which a perturbation transforms a dynamically stable activity into unintended interacting changes of states with a harmful outcome."

For McClay (2): ". . . a loss incident is an unintentional, unexpected, occurrence (resulting from uncontrolled hazards) which—without any subsequent events—has the potential to produce damaging and injurious effects suddenly or, if repetitive, over a long period of time."

In MORT concepts, unwanted energy releases and hazardous environmental conditions are significant causal factors. McClay approaches the subject of unwanted energy releases and introduces "mass" as an element in this manner (2).

> It can be deduced and has also been empirically shown, that the actual damaging and injurious effects are produced by the release, transformation, or misapplication of energy. Since mass and energy are interconvertible, the release, transfer, or misapplication of mass should also be seen as having the potential to produce these same adverse effects.

If there are no unwanted releases or transfers of energy, or exposures to hazardous environments, hazards-related incidents will not occur. That statement applies to *every* incident category: overexertion (back injuries, strains, and sprains}; slips; falls; caught in between; struck by an object; electrocution; reaction to a chemical; et cetera.

A HAZRIN causation model, then, should encompass unwanted energy releases and unwanted exposures to hazardous materials within the incident process.

Dr. William Haddon, Jr. (19, 20) was the first director of the National Highway Safety Bureau. He is credited with having developed the energy release concept. His thinking was that an unwanted energy release or an exposure to a hazardous environment can be harmful and that a systematic approach to limiting such a possibility should be undertaken. That is the framework on which The Management Oversight and Risk Tree—MORT is built. Having knowledge of Haddon's energy release concepts augments professional safety practice.

Benner (17, 18), Ferry (4), Johnson (3), and McClay (2) are valuable resources concerning an understanding of the hazards-related incident phenomenon. These excerpts from their writings give an indication of their thinking.

For Benner (18):

> An accident occurs in connection with an activity involving certain interrelated elements. These activities are conducted in the presence of conditions of vulnerability (interrelated with each other or the activity) which conditions must exist for an accident to be possible. An accident begins when one of the elements engaged in the activity is overtaxed beyond its ability to recover from the overload and cannot resume functioning within the limits of its capability again in the continuity of the activity.
>
> The accident process can be described in terms of specific interacting actors [actors may be things or people], each acting in a separate and spatial logical relationship. By breaking down the events seen as the accident into increasingly more definitive sub-events, the understanding of the phenomenon increases with each successive breakdown.

Benner stresses that interrelated and multiple events proceed on a time scale, displayed much like a musical score, and culminate in some outcome. Events in the sequence may occur on a single track or impact on other events on parallel tracks. Events sequences are to be plotted on parallel lines along a time scale, in accord with their place in the sequence.

In several of Benner's writings, appropriate recognition is given to MORT as a foundational thought process. Johnson, in *MORT Safety Assurance Systems* (3), gives recognition to Benner's research. For both, charting systems are used to graphically depict events contributing to a hazards-related incident, from the beginning to the end of the process.

McClay's approach (2) to causation is somewhat different but also has some similarity to Benner's postulations and the bases on which MORT is founded. McClay indicates that:

> . . . not all factors which contribute to a loss incident occur at the same time just before the incident occurs . . . causal factors can be placed into two groups with respect to the temporal nature of their occurrence . . . causal factors which exist or occur within the same specific time frame and location as the loss incident (called proximate factors) . . . and causal factors which do not exist or occur within the same specific time frame and general location as the loss incident (called distal factors).

McClay's premise, as I interpret it, is that the "loss incident" occurs over time and that there may be a variety of causal factors involved. Thus, there is some continuity in what McClay, Benner, and Johnson have written.

Ferry, Benner, and Johnson stress the complexity of HAZRINS. Ferry (4) writes that:

> It has been found that simple mishaps tend to be complex in terms of many causal factors with lengthy sequences of errors and changes leading to the various events. This makes it essential for the investigator to have a system, a methodology for breaking down the entire sequence of events into individual events with supporting information.

Benner (18) says:

> It is the accident process which is complex and requires a detailed comprehension before it can be understood adequately to control the inherent safety problems.

In Johnson (3):

> Accidents are usually multifactorial and develop through relatively lengthy sequences of changes and errors. Even in a relatively well-controlled work environment, the most serious events involve numerous error and change sequences, in series and parallel.

Building on the foregoing, this definition of a hazards-related incident is now proposed:

> A hazards-related incident, a HAZRIN, is an unplanned, unexpected process of multiple and interacting events, deriving from the realization of uncon-

trolled hazards and occurring in sequence or in parallel, which is likely to result in harm or damage.

Causal Factors Derive Principally from Upstream Decisions

To prevent the initiation of these complex, multifactorial sequences that may result in harm or damage, "intervention" is proposed by many authors, very often concentrating on the identified unsafe act or condition. I suggest that to be most effective, intervention has to impact on the beginnings of things, on the upstream decision-making out of which a system is created.

An operating system, whatever the product made or the service provided, is a reflection of an organization's culture, its values, and its sense of responsibility to its employees and its community, all of which determine its design decisions for hardware, facilities, and the work environment; the work methods; and the management aspects of operations.

Management policies, organization, and commitment are shown together as one element in some causation models. One could argue that management commitment and involvement deserve separate and distinct consideration since they are the reflection and extension of the organization's culture, from which all hazard prevention and control decisions derive.

An organization's culture and its management commitment should stand as separate and sequential items in a causation model. Aspects of Less Than Adequate performance in the design aspects, the management and operational aspects, and the task performance aspects of safety derive from the culture and the management commitment.

First, evidence of an organization's culture and management's commitment with respect to safety is displayed through its design and engineering decisions. Where hazards are given the required consideration in the design and engineering processes, a foundation is established that gives good probability to avoiding hazards-related incidents.

Conclusion

Just what observations can be made from all this? If safety professionals choose to examine their causation models, which are the basis of the advice they give, this thought process is proposed for consideration, as a beginning:

1. Professional safety practice requires that the advice given be based on a sound hazards-related incident causation model so that, through the application of that advice, hazards are effectively avoided, eliminated, or controlled.

2. A hazards-related incident, a HAZRIN, is an unplanned, unexpected process of multiple and interacting events, deriving from the realization of uncontrolled hazards and occurring in sequence or in parallel, that is likely to result in harm or damage.
3. Hazards-related incidents, even the ordinary and frequent, are complex as respects the multiple and interacting causal factors that may contribute to them.
4. Unwanted releases of energy and exposures to hazardous environments are fundamental in the occurrence process for hazards-related incidents.
5. Causal factors include all elements of technology or activity—the characteristics of things, the actions or inactions of people, and the events that contribute to the incident process.
6. A HAZRIN causation model should:
   - recognize that an organization's culture is the primary influence concerning hazards development;
   - give separate recognition to management commitment as the extension of the organization's culture, and as the source of the decision-making affecting the avoidance, elimination, and control of hazards;
   - move the emphasis of the practice of safety to the origins of decision-making;
   - emphasize that a balance of considerations is needed for causal factors deriving from less than adequate policies, standards, or procedures that impact on the design management, operations management, and task performance aspects of safety.

It is a necessity that the advice given by safety professionals be based on an understanding of the reality of causal factors and actually serves to attain a state for which the risks are judged to be acceptable. I shall try to develop a systemic causation model for hazards-related occupational incidents that represents the thoughts set forth in this essay.

REFERENCES

1. Benner, Jr., Ludwig. "Rating Accident Models and Investigative Methodologies," *Journal of Safety Research* 16, no. 3 (Fall 1985).
2. McClay, Robert E. "Toward a More Universal Model of Loss Incident Causation," *Professional Safety* (Jan/Feb 1989).
3. Johnson, William G. *MORT Safety Assurance Systems.* New York: Marcel Dekker, 1980.
4. Ferry, Ted S. *Modern Accident Investigation and Analysis.* New York: John Wiley & Sons, 1981.
5. Benner, Jr., Ludwig. "Accident Investigations: Multilinear Sequencing Methods," *Journal of Safety Research* (June 1975).

6. Fischhoff, Baruch. *Risk: A Guide To Controversy. Appendix C: Improving Risk Communication.* Washington, D.C.: National Academy Press, 1989.
7. Deming, W. Edwards. *Out of the Crisis.* Cambridge, MA: Center for Advanced Engineering Study, Massachusetts Institute of Technology, 1986.
8. Heinrich, H. W. *Industrial Accident Prevention,* 3rd ed. New York: McGraw-Hill, 1950.
9. Johnson, William G. *The Management Oversight and Risk Tree—MORT.* Washington, D.C.: U.S. Government Printing Office, 1973.
10. Mirer, Franklin E. "Safety and Health Program Management Standard: The Need For a Performance Oriented Approach That Incorporates the Principles of the Quality Movement." A speech given at an American National Standards Institute (ANSI) workshop, May 1996.
11. *Guide to Use of the Management Oversight and Risk Tree.* SSDC-103. U.S. Department of Energy, Office of Safety and Quality Assurance, Idaho National Engineering Laboratory, Idaho Falls, Idaho, 1994.
12. Petersen, Dan. *Techniques of Safety Management.* Goshen, NY: Aloray, 1989.
13. Chapanis, Alphonse. "The Error-Provocative Situation." In *The Measurement of Safety Performance,* ed. William E. Tarrants. New York: Garland Publishing, 1980.
14. Kletz, Trevor. *An Engineer's View of Human Error.* Rugby, Warwickshire, UK: Institution of Chemical Engineers, 1991.
15. Paradies, Mark. "Root Cause Analysis and Human Factors," *Human Factors Society Bulletin* (August 1991).
16. DeJoy, David M. "Toward a Comprehensive Human Factors Model of Workplace Accident Causation," *Professional Safety* (May 1990).
17. Hendrick, Kingsley, and Ludwig Benner, Jr. *Investigating Accidents with STEP.* New York: Marcel Dekker, 1987.
18. Benner, Jr., Ludwig. "5 Accident Perceptions: Their Implications For Accident Investigators," *Hazard Prevention* (Sept/Oct 1980).
19. Haddon, Jr., William J. *Preventive Medicine: The Prevention of Accidents.* Boston: 1966.
20. Haddon, Jr., William J. "On the Escape of Tigers: An Ecological Note," *Technology Review* (May 1970).

Chapter 7

A Systemic Causation Model for Hazards-Related Occupational Incidents

Introduction

This causation model for hazards-related occupational incidents recognizes:

- the impact an organization's culture, reflected by its policies, standards, procedures, and accountability systems, and their implementation, has on causal factor development; and
- the need for a balanced, systemic approach that appropriately addresses design management, operations management, and task performance causal factors.

This model promotes inquiry to determine the upstream sources from which causal factors originate. Its use requires applying basic problem-solving techniques to identify actual causal factors. Whether an organization has two or 2 million employees, the model is applicable.

Most causation models have minimized less than adequate design and engineering concepts and outcomes as a source of causal factors for hazards-related incidents, with one significant exception. That exception is MORT—The Management Oversight and Risk Tree (1, 3). Concepts on which MORT is based have influenced my thinking greatly, and I am indebted to all who worked on the creation and betterment of MORT.

Unfortunately, the causation model I suggest cannot be presented simply. As incident causation can be complex, so also must a model depicting the incident phenomenon be somewhat complex. Nevertheless, I hope that the ideas on which the model depicted in Figure 7.1 is based are presented clearly.

Figure 7.1 A Systemic Causation Model for Hazards-Related Occupational Incidents

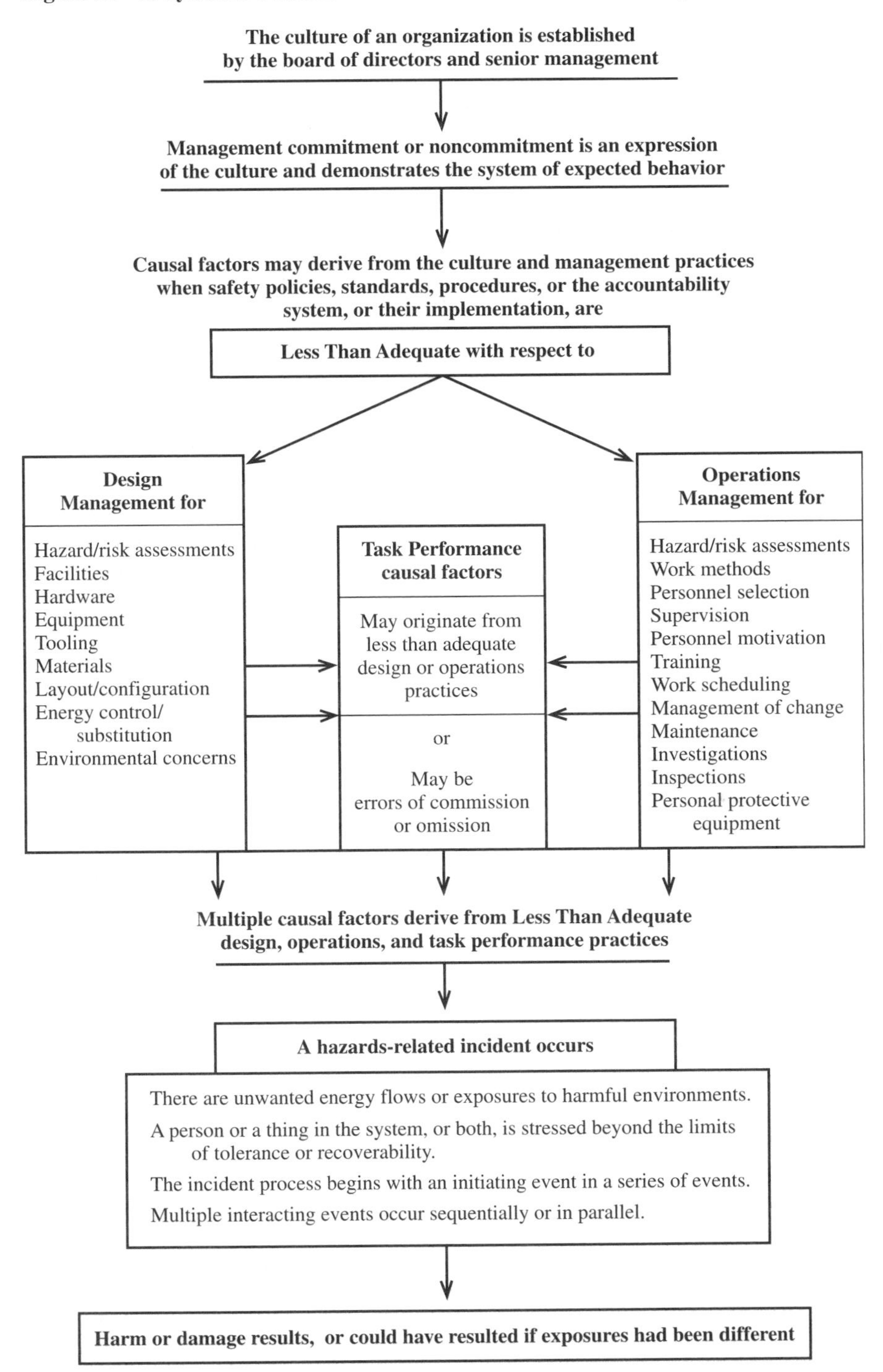

Certain assumptions were made in the development of this hazards-related incident causation model, which:

1. Recognizes that an organization's culture is the primary influence concerning the existence of hazards;
2. Gives separate recognition to management commitment or noncommitment to safety as an extension of an organization's culture, and as the source of the management decision-making effecting the avoidance, elimination, or control of hazards;
3. Moves the emphasis of the practice of safety to the origins of decision-making, rather than on outcomes;
4. Emphasizes that a reasoned approach is needed for the identification of causal factors deriving from *Less Than Adequate* policies, standards, procedures, and accountability systems, or their implementation practices, which impact on the aspects of safety, encompassing
   (a) design management,
   (b) operations management, and
   (c) task performance;
5. Recognizes unwanted energy releases or exposures to hazardous environments as necessary to the occurrence of hazards-related incidents;
6. Establishes that hazards-related incidents, even the ordinary and frequent, are complex and will have multiple and interacting causal factors contributing to their occurrence, acting sequentially or in parallel.

DEFINITIONS

To promote an understanding of this systemic causation model, meanings are given here of the terms *incidents, hazards, HAZRINS, model, causation, causal, factors,* and *causal factors.*

Incidents as a term encompasses all hazards-related events (HAZRINS) that have been referred to as accidents, mishaps, near-misses, occupational illnesses, environmental spills, losses, fires, explosions, et cetera. All of the incidents to which this causation model applies derive from hazards. There are no exceptions.

Hazards include all aspects of technology and activity that produce risk. A *hazard* is defined as the potential for harm: Hazards include the characteristics of things and the actions or inactions of people.

A hazards-related incident, a *HAZRIN,* is an unplanned, unexpected process of multiple and interacting events, deriving from the realization of uncontrolled hazards and occurring in sequence or in parallel, that is likely to result in harm or damage.

A *model* is to represent the theoretical ideal for the process through which hazards-related incidents occur, a process that requires determining when the phenomenon begins and ends.

Causation means the act or agency of causing or producing an effect. *Causal* means of, constituting, or implying a cause. *Factor* means one of the elements contributing to a particular result or situation. *Causal factors* include all of the elements—the events, the characteristics of things, and the actions and inactions of people—that contribute to the incident process.

PREMISES ON WHICH THIS CAUSATION MODEL IS BASED

A. An organization's culture determines the level of safety attained.

B. Management commitment or noncommitment to safety is an expression of an organization's culture, an expression of its system of expected behavior.

C. Policies, standards, and procedures concerning safety, and their implementation and accountability systems—all of which derive from the organization's culture and the management commitment or noncommitment to safety—MAY BE:

1. *Less Than Adequate* with respect to hazard analyses and risk assessment processes, for the identification, avoidance, elimination, and control of hazards.
2. *Less Than Adequate* for Design and Engineering Management concerning:
   - Facilities
   - Hardware
   - Equipment
   - Tooling
   - Materials
   - Layout and configuration
   - Energy control/substitution
   - Environmental concerns
3. *Less Than Adequate* for Operations Management concerning:
   - Work methods
   - Personnel selection
   - Supervision
   - Personnel motivation
   - Training
   - Work scheduling
   - Management of change
   - Maintenance
   - Investigations
   - Inspections
   - Personal protective equipment

D. *Less Than Adequate* design practices may be the direct source of incident causal factors, or may lead to task performance causal factors.

E. *Less Than Adequate* operations practices may be the direct source of incident causal factors, or may lead to task performance causal factors.

F. *Less Than Adequate* task performance may be the direct source of causal factors, consisting of:

- Errors of commission; or
- Errors of omission.

G. Because multiple causal factors derive from *Less Than Adequate* design management *and* operations management *and* task performance errors, a hazards-related incident occurs.
H. Harm or damage results from the incident, or could have resulted if the exposures had been different.

Supporting Discussion

In this supporting discussion, the influence of system safety concepts and the concepts on which MORT—The Management Oversight and Risk Tree is based will be evident.

Organizational Culture

An organization's culture consists of its values, beliefs, legends, rituals, mission, goals, performance measures, and its sense of responsibility to its employees, customers, and community, all of which are translated into a system of expected behavior.

The board of directors and the senior management obtain, as a derivation of the organization's culture, the hazards-related incident experience that they establish as acceptable. For the personnel in the organization, what is "acceptable" is their interpretation of the reality of what management does, which may differ from what management says.

When what is acceptable to the board of directors and to senior management concerning safety is *Less Than Adequate* in relation to the risks faced by the organization's employees, causal factors for incidents may derive, then, from the organization's culture.

Management Commitment or Noncommitment to Safety—An Extension of the Organization's Culture

An organization's culture is translated into a system of expected behavior that is expressed by the level of management commitment or noncommitment to safety. All aspects of safety, favorable or unfavorable, derive from that commitment or noncommitment.

Where management commitment to safety is solid, management achieves an understanding through its actions that all in the organization are to manage their endeavors with respect to hazards so that the risks deriving from those hazards are acceptable.

Causal factors may originate from *Less Than Adequate* management safety policies, standards, procedures, and accountability systems, or their implementation.

Hazard Analysis and Risk Assessment

Using hazard analysis and risk assessment methods is vital in achieving an acceptable risk level for the design, the operations, and the task performance aspects of safety.

A hazard analysis is a process that produces responses to these questions: What are the characteristics of things and the actions and inactions of people that present a potential for harm? Can the potential be realized? Who and what are exposed to harm or damage? What is the frequency of endangerment? What will be the severity of harm or damage if the potential is realized?

A risk assessment is an analysis that addresses both the probability of a hazards-related incident occurring, and the expected severity of harm or damage that may result.

Effective application of hazard analysis and risk assessment methods results in hazards being identified, avoided, eliminated, or controlled in the design or redesign processes.

In *MORT Safety Assurance Systems* (1), William G. Johnson was correct when he wrote:

> Hazard identification is the most important safety process in that, if it fails, all other processes are likely to be ineffective.

Causal factors resulting from *Less Than Adequate* hazards analyses and risk assessments must be recognized for their significance. For new facilities or modified systems, such inadequacies result in hazards being brought into the workplace inadvertently. For existing operations, failure to identify causal factors when task reviews are made or incidents are investigated results in ineffective hazards management.

Design Management

Strong emphasis is given in this causation model to the causal factors that arise out of *Less Than Adequate* design management practices. As the term is used here, design encompasses all processes applied in devising a system to achieve results.

An organization's culture regarding safety is outwardly demonstrated through its initial design decisions and subsequent redesign decisions that determine what the facilities, hardware, equipment, tooling, materials, layout and configuration, and the work environment are to be.

I admit to being a disciple of W. Edwards Deming, who is world-renowned in quality management. My interpretation of Deming is that he stressed again and again in *Out of the Crisis* (2) that processes must be designed, or redesigned, to achieve superior quality if such performance is desired, and that superior quality cannot be attained otherwise. The same principle applies to safety.

Too much emphasis cannot be given to the causal factors that derive from *Less Than Adequate* design decisions. If hazards are not adequately avoided, eliminated, or controlled in the design processes, multiple causal factors will originate from those inadequacies.

Operations Management

This causation model puts a major focus on *Less Than Adequate* management practices that impact on the operations system as a source from which causal factors derive.

A significant excerpt is taken from the *Guide to Use of the Management Oversight and Risk Tree* (3) to emphasize focusing on the system:

> It should be pointed out that the kinds of questions raised by MORT are directed at systematic and procedural problems. The experience, to date, shows that there are few "unsafe acts" in the sense of blameful work level employee failures. Assignment of "unsafe act" responsibility to a work level employee should not be made unless or until the preventive steps of: (1) hazard analysis, (2) management or supervisory detection, and (3) procedures safety reviews have been shown to be adequate.

Surely, employees should be trained and empowered up to their capabilities, and procedures should be established for employees to make contributions to safety. But employees should not be expected to do what they cannot do. Nor should the focus be on their behavior (the so-called unsafe act) when the causal factors for hazards-related incidents derive principally from *Less Than Adequate* design or operations management. Employees are greatly limited by the work system—established by, and under the control of, management.

While the focus is on the operations system, created by management and which only management can change, it is not intended that the significance of employee-centered, task performance causal factors be diminished.

So, causal factors originate when operations practices are *Less Than Adequate* concerning work methods; personnel selection; supervision; personnel motivation; training; work scheduling; management of change; maintenance; investigations; inspections; and personal protective equipment.

Relationships: Design Decisions, Operations Practices, and Task Performance

There must be a recognition of the impact design decisions that determine the physical workplace and the prescribed management and operations practices have on task performance.

That premise reflects the work of Dr. R. J. Nertney, who developed the "Nertney Hazard Analysis Wheel" (4), which is:

> . . . a simple, provocative method of examining the successive phases of hardware-procedure-personnel development and also examining the all important interfaces between those three elements.

Think about it: work methods prescribed to get a job done are determined largely by the design decisions made concerning facilities, hardware, equipment, and layout and configuration, et cetera. That applies whatever the complexity of the tasks—cutting an invoice or making airplanes. Standards for the selection of personnel and their training must match the "hardware" decisions.

When the match of the design of facilities, hardware, equipment, tooling, et cetera, and operations methods, and personnel practices (selection, training, et cetera) is *Less Than Adequate,* causal factors attributable to the mismatch will arise.

Also, while *Less Than Adequate* design decisions and *Less Than Adequate* operations practices may be the direct source of incident causal factors, they may also lead to task performance causal factors. Causal factors will derive directly from design decisions or operations practices if the design of the workplace, or the work methods:

- is overly stressful
- induces fatigue
- is error-provocative
- promotes riskier behavior than prescribed work methods
- is unnecessarily difficult or unpleasant
- is unnecessarily dangerous
- requires "jerry rigging" for job accomplishment
- does not allow easy access for the work to be done

Because Multiple Causal Factors Develop a Hazards-Related Incident Occurs

It is the exceptional hazards-related occupational incident that does not have multiple causal factors. This is the process as causal factors develop and an incident occurs:

- There are unwanted energy flows or exposures to hazardous environments. (If there are no unwanted energy flows or exposures to hazardous environments no incidents can occur that might result in harm or damage.)
- A person or thing in the system, or both, is stressed beyond the limits of tolerance or recoverability.
- The incident process begins with an initiating event in a series of events.
- As the incident process continues, multiple and interacting events occur, sequentially or in parallel.

The series of events comes to an end. A hazards-related incident, then, is not a single event, but a group of dynamic actions, or sets of interacting or interconnecting events.

Incident Outcome

Harm or damage to people, property, or the environment results from the incident, or could have resulted if the exposures had been different. The intent in the wording used here is to give emphasis to incidents that have a potential for adverse effects rather than just those that do result in harm or damage.

References

1. Johnson, William G. *MORT Safety Assurance Systems.* New York: Marcel Dekker, 1980.
2. Deming, W. Edwards. *Out of the Crisis.* Cambridge, MA: Center for Advanced Engineering Study, Massachusetts Institute of Technology, 1986.
3. *Guide to Use of the Management Oversight and Risk Tree.* SSDC-103. U.S. Department of Energy, Office of Safety and Quality Assurance, Idaho National Engineering Laboratory, Idaho Falls, Idaho, 1994.
4. Nertney, R. J. "Nertney Hazard Analysis Wheel." In *MORT Safety Assurance Systems.* William G. Johnson. New York: Marcel Dekker, 1980.

Chapter 8

Incident Investigation: Studies of Quality

Introduction

Why a study of the quality of incident investigation? There are three major elements in practice of safety: pre-operational (in the design processes); in the operation mode (integrated within a process of continuous improvement); and post-incident (after a hazards-related incident has occurred).

In this third and important element, competent investigation of hazards-related incidents is vital. Effective safety practice requires that *actual causal factors*—the hazards and events that contributed to the incident process—be identified, evaluated, and eliminated or controlled.

Incident investigation, done well or superficially, reflects the reality of an organization's culture concerning safety. The quality of incident investigation gives significant messages to employees as they interpret the substance of what management does in relation to what management says.

Thorough incident investigation and follow-through with remedial actions support a culture that gives importance to safety. Poorly done incident investigations give employees reasons to doubt management's sincerity with respect to safety.

Hazard Analysis Is the Most Important Safety Process

This statement, with which I emphatically agree, comes from *MORT Safety Assurance Systems* (1) by William G. Johnson.

Hazard analysis is the most important safety process in that, if that fails, all other processes are likely to be ineffective.

If hazard identification and analysis do not relate to actual causal factors, the resulting corrective actions proposed will be misdirected, and ineffective. Superior incident investigations are required to identify and evaluate actual causal factors so that appropriate corrective actions can be taken.

Data Gathered for the First of Two Studies

To study the quality of incident investigation as actually performed, a collection was made of reports completed by supervisors and investigation teams after injuries and illnesses had been reported. From thirty-seven locations of eleven organizations, using fifteen different forms, 537 investigation reports were collected and reviewed. (With emphasis, I state that the study I made would not meet the modeling and methods requirements of scientific inquiry.)

Highlights of the Study

General observations deriving from the study follow. They are just that—observations resulting from a subjective review of a limited number of reports, the 537 reports received. I do suggest that these observations deserve a broader consideration through a study that would meet the requirements of good science since the subject—incident investigation—is so important.

1. Using a scale for quality of investigations of 1 to 10, with 10 being best, organization scores ranged from a high of 8 to a low of 2. For the entire study, 5.5 was the weighted overall score.
2. Investigation of incidents is done well in those entities where the culture includes a management accountability for hazards-related results.
3. Variations in structure and content requirements of incident investigation forms were extensive.
4. Good incident investigation cannot be achieved without training, and repeated training.
5. Where supervisors had participated in job hazard analyses, they seemed to have a better understanding of causal factors and did a more thorough job of incident investigation.
6. General information entries (name, social security number, occupation, et cetera) and incident descriptions were usually the most complete parts of reports.

7. Incident report forms collected are predominantly "Heinrichian": they emphasize "man failure," the so-called unsafe acts, as the principal causal factor. "Man failure" is a term significant in Heinrich's premises.
8. Although not necessarily intended, some report formats lead to identification of a single causal factor, emphasizing unsafe acts, rather than stressing the concept of multiple causation. Sometimes—but only sometimes—a contributory causal factor was also identified.
9. It is probable that supervisors, who make only one, two, or three incident investigations a year, don't fully understand the descriptions of the various causal categories given in investigation forms.
10. Form content requirements seldom lead to examinations of the design of the workplace or the design of work methods.
11. For approximately 38 percent of the reports reviewed, entries suggested that further inquiry should have been made respecting design of the workplace or of the work methods.
12. In the one instance reviewed in detail, cause codes entered for subsequent computer analysis did not match the contents of the investigation reports.
13. Injury type and incident type codes were close to reality.
14. For many reports, plausible causal factors could not be identified.

Some organizations are doing a very good job of incident investigation. Reports received from them were commendable, proving that effective incident investigation can be achieved. In other companies, what is accepted indicates inadequate knowledge by those completing investigations, and an absence of management accountability for the quality of incident investigation. Completion of reports in those instances is obviously a perfunctory exercise, with little value.

Structure and content variations in incident investigation forms were extensive, and precise and equivalent evaluations could not be made in report reviews. As an example, one of the simplest forms consisted in its entirety of these questions. What happened? Why did it happen? What should be done? What have you done so far? How will this improve operations?

Methodology and Scoring

Using subjective judgments, scorings of these individual subjects were first given where possible:

- general information
- incident descriptions
- causal factors determinations
- corrective actions proposed

Then, the entire incident report was scored. Subsequently, a composite score for the entity's incident investigation system as a whole was computed.

A quality score of 10 was best, using a 1 to 10 scale. All scores were rounded to whole numbers. Composite scores ranged from a high of 8 to a low of 2, with an overall weighted score of 5.5 for the entire study. Following is the range of scoring.

Table 8.1. Composite Scores

Individual	10	9	8	7	6	5	4	3	2	1
Entity			x	x	x	x		x	x	
Scores			x	x		x		x	x	
Overall Weighted Score 5.5										

Comments by Safety Professionals

Discussions of achievements with safety professionals whose organizations had top scores did not produce any surprises. Incident investigation for hazard identification and analysis gets done best where the organization's culture includes an accountability for superior performance. An aggregate list follows of the comments made in discussions with safety professionals in those entities with the best incident investigation systems.

- A focus on safety in those entities comes from the top, as a reflection of the organization's culture.
- If incident investigation is important to the boss, if it is a part of the accountability system, it will be done well. Where that is the case, safety professionals will receive requests for help from all levels of management.
- For everyone from the senior vice president down to the floor supervisor, the annual performance review includes safety results.
- Safety performance is an element scored in the bonus program.
- There is an annual competition on safety performance; statistics produced monthly are treated seriously.
- Only one company award has the president's name on it—the safety achievement award. Presentation of the award is made by the president at a significant social event.
- For every recordable incident, the location manager is required to submit a report to the group vice president.
- At all manufacturing and warehousing locations, there is a Job Hazard Analysis for every job.

- Supervisory personnel cannot do a good job of investigation if they have not had the necessary training. Thus, incident investigation training is given, and repeated.

Report Titles

Incident investigation reports received had a variety of titles, although some were identical. They come under these names, which in themselves have significance:

- Investigation and Findings of Injury/Illness
- Supervisors Report of Accident Investigation
- Mishap Report
- Accident Investigation Report
- Injury/Illness Report
- Incident Investigation Report: Personal Investigation
- Supervisor's Investigation Report
- Occupational Injuries or Illnesses Report

General Information Required

General information entries were usually well completed. A summary of entries required by all forms include:

Name
Sex
Shift/time
Date of birth
Department/division/sector
Seniority date/company service
Date of accident/illness
Where sent/hospital/home/other
Did employee die
Contributory cause code
Injury/illness code
Was place of accident on employer's premises
Social security number
Clock number
Marital status
Address
Occupation/job title
Number of years/months on job
To whom was accident reported
Describe medical treatment
Basic cause code
Accident code
Place of accident

Incident Descriptions

Incident descriptions were often the most complete parts of reports. In many instances, incident descriptions were the only complete parts of reports. It is easier, obviously, to describe what happened than it is to determine causal factors and what corrective actions should be taken.

Quality of Causal Factor Determination

Variations in the quality of causal factor determination were extreme. Of the fifteen forms received, ten direct the person who completes the form to first identify the unsafe act of the employee. That requires seeking evidence of the "man failure" foundational in "Heinrichian" premises.

After a review of his theorems developed in the 1920s and illustrated by the "domino sequence," H. W. Heinrich wrote this in the third edition of *Industrial Accident Prevention* (2):

> From this sequence of steps in the occurrence of accidental injury it is apparent that man failure is the heart of the problem. Equally apparent is the conclusion that methods of control must be directed toward man failure.

It is a prominent practice, whether intended or not, to put the principal responsibility (not blame) for the incident on something the employee did or did not do—that is, to seek evidence of "man failure." Some of the forms also asked that unsafe conditions be recorded. Sometimes, but seldom, they contain references to design and systems shortcomings.

In those organizations where incident investigation is done well, supervisors who obviously had training in incident causation and who were familiar with job hazard analysis procedures would frequently go beyond the form's requirements. It was not unusual for those supervisors to pose questions about and seek help on the design of the workplace and the design of work methods. It regularly occurred in those organizations that supervisors would record on incident investigation reports something like "this job needs a new job hazard analysis."

For about 38 percent of the reports received, incident descriptions and discussions of causal factors suggested that further inquiry should have been made respecting design of the workplace or design of work methods. In far too many cases where recordings indicated that a review of workplace design or work methods design might be beneficial, unsafe acts of employees were selected as the primary incident causes. And the corrective actions proposed were to obtain behavior modification.

Of the fifteen variations of incident investigation reports received, six promote an overly simplistic and inappropriate approach to causal factor determination. They reflect this instruction, although somewhat ancient, given in a publication of the American National Standards Institute (3).

> It is recognized that the occurrence of an injury frequently is the culmination of a sequence of related events, and that a variety of conditions or circumstances may contribute to the occurrence of a single accident. A record

> of all these items unquestionably would be useful to the accident preventionist.
>
> Any attempt to include all subsidiary or related facts about each accident in the statistical record, however, would complicate the procedure to the point of impracticality. The procedure, therefore, provides for recording only one pertinent fact about each accident in each of the specific categories or classifications.

Usually, there are multiple causal factors for hazards-related incidents. Yet there seems to be a desire to retain an age-old theme of simplicity. If people who make incident investigations are directed to select "one pertinent fact . . . . in each of the specific categories," more than likely that is what they will do. And the value of the investigation is diminished. Where incident investigation is done best, multiple causal factors are sought, and it is the exception when only a single causal factor is recorded.

Generally, few supervisors did a good job of ergonomics problem identification. Although there has been an emphasis on ergonomics as a significant aspect of the practice of safety, that emphasis was not seen in the content of the incident investigation reports reviewed.

It is a too common practice that personnel with "safety" in their titles place their signatures, indicating acceptance and approval, on incident investigation reports that are far from adequate, particularly concerning causal factor determination.

Surprisingly, the terms "careless" or "carelessness" or "should have been more careful" appeared as causal factors only seven times in 537 reports.

Although "employee empowerment" has been prominent in contemporary management literature, there were no indications in this study that personnel other than supervisors or investigation teams were completing incident investigation reports.

Cause Categories Used in Forms May Not Be Understood

Terms such as *basic cause, contributory cause, immediate cause,* and *management cause* appear in the forms. Definitions on the forms were understandably brief; some were confusing. I was not always certain of their meanings. Entries on the forms indicate that supervisors, who may complete one, two, or three incident investigation reports a year, do not understand what the terms mean.

Causal Factor Codes May Not Match Report Content

Five of the fifteen forms received require entry of causal factor, incident type, and injury type codes. I have been an aggressive promoter of the computer entry of causal factor codes for later analysis. Now, I pose a question about the accuracy of the causal data that can be derived from investigation reports completed by supervisors.

For a group of 121 reports, these basic cause codes were entered in the boxes for such codes, all of them employee-action related:

28% Failure to follow established procedure
27% Haste, inattention, shortcut
6% Improper use of equipment, tools, or materials

61%

Basic cause codes selected did not match the written entries on the forms. In the entities that submitted those reports, supervisors were doing a better job of identifying possible causal factors than the codings indicated. Corrective actions taken were in greater depth than would be required by the cause code entries. Computer runs would identify, erroneously, the principle causal factors based on the cause code entries. Making an assumption that the data is factual would result in a misdirection of efforts.

Safety professionals who have instituted systems requiring the entry of cause codes on investigation reports for which computer summaries are produced might want to conduct a similar exercise to determine whether the computer output matches reality. I suggest a close look at the causal data being obtained, and a realistic assessment of what can be achieved.

Injury Type and Incident Type Codes

Injury type and incident type codings were at a much higher accuracy level than codes for causal factors. Incident type and injury type analyses would provide information from which to commence further inquiry.

Investigations by Teams Were Superior

In some entities, it is required that an incident investigation team be selected and gathered if the results of the incident were serious or could have been serious under other circumstances. Reports prepared by investigation teams (twelve of them were received) were a pleasure to read. Each one reflected an understanding of multiple causation and pursued several routes in causal factors determination and in selecting corrective actions.

OBSERVATIONS FROM A SECOND STUDY

Later, while researching for "Measurement of Safety Performance" (Chapter 20) I observed that companies with superior OSHA rates also had lower workers compensation costs in comparison to other companies. I then wanted to determine whether the quality of investigation of hazards-related incidents would also be superior. (For this second study, it must also be said that the methodology used would not stand the test of good science.)

The public perception is that the operations of the industry I chose for this inquiry, the chemical industry, are high hazard even though its average OSHA rate is commendably low. For the six companies that provided the 328 investigation reports I reviewed, the range of their OSHA Recordable Rates is 1.6 to 2.9. For their Lost Work Day Case Rates, the range is 0.6 to 1.3. Comparable rates for private industry, overall, are 8.5 for OSHA Recordables and 3.8 for Lost Work Day Cases.

With just a bit of refinement, the scoring system used for the quality of investigations was the same as that used in the first study, a scale of 1 to 10, with 10 being highest. Entity scores in the second study ranged from a low of 7.4 to a high of 8.7. Entity scores in the original study had a low of 2 and a high of 8. Overall, the weighted score was 8.1 in the second study; it was 5.5 in the first study. Obviously, outcomes were significantly different.

In each of the six companies, a sound safety culture has existed for several years. Management personnel at all levels are held accountable for hazards-related results.

While very few of the incident reports collected in the first study were completed by investigation teams, every report submitted in this second study by three companies was a team effort. In the other three companies, teams are assigned to investigate every OSHA recordable incident.

Since that seemed a bit much in relation to what is done elsewhere, I asked a safety director how the activity was supported in these times of lean staffs. His answer: "We know that the time expended may seem excessive, but the procedure gets a lot of people involved and that reinforces their belief that in our company safety is truly important. Thorough incident investigation is a part of our culture."

For almost all reports, causal factors identified were plausible, with a good balance between consideration of design and engineering practices, operation practices, and task performance. Citing multiple causal factors was prevalent. References to written job procedures deriving from job hazard analyses or other job study methods were frequent. As in the original study, supervisors who had participated in job hazard analyses had a better understanding of causal factors and did a more effective job of investigation.

Seldom was there a mention of ergonomics-related causal factors in the reports collected for the original study. In the second study, extensive knowledge of ergonomics was displayed. Obviously, the ergonomics awareness and training initiatives undertaken in the companies participating in this second study have had an effect.

While this second study cannot be considered conclusive, it is probable that a further and more scientific study would establish that effective incident investigation reaps many benefits—as a productive means of eliminating or controlling hazards and achieving fewer injuries and illnesses, as an augmenting factor in attaining reduced workers compensation costs, and as a supporting element within a sound safety culture.

References

1. Johnson, William G. *MORT Safety Assurance Systems.* New York: Marcel Dekker, 1980.
2. Heinrich, H. W. *Industrial Accident Prevention,* 3rd ed. New York: McGraw-Hill, 1950.
3. *Method of Recording Basic Facts Relating to the Nature and Occurrence of Work Injuries—Z16.2.* New York: American National Standards Institute.

Chapter 9

Designer Incident Investigation

Introduction

It would be folly to suggest that an incident investigation system could be drafted that would universally apply in all organizations. This essay is to serve as a reference for those who are to design, or redesign and improve, an incident investigation system, taking into consideration

- the very large majority of entities that have not or cannot achieve a level of sophistication that would be expected if all investigations were made by safety professionals or trained investigation teams, and
- what can realistically be expected of line supervisors, middle managers, and location managers in such organizations.

More specifically, this essay will assist a safety professional in crafting an incident investigation instructional procedure, an investigation form or outline, a refresher guide for causal factor and corrective action determination, and a training program.

In determining what incident investigation system is to be adopted, an assessment should be made of what is practicably attainable. That requires making assumptions about the organization in which a safety professional resides. If there is little management support for incident investigation, or if training cannot be given to those who are to make investigations, it would be best to opt for simplicity. But including a modest stretch goal in the system would be a good idea.

Reality Observations

Effective investigation and analysis of hazards-related incidents are important elements in a quest to achieve superior results in safety. A high quality of incident investigation and analysis results not only in risk reduction, but also serves as a positive reinforcement for the safety policies and practices management has established.

As two studies have shown (see Chapter 8, "Incident Investigation: Studies of Quality"), the effectiveness of incident investigation varies greatly among organizations. To a large extent, the variations reflect the differences in organizational culture, the level of safety performance that is established by management policies and practices as being acceptable, and the resources made available.

Understandably, the several texts that include principles and practices to be applied in incident investigation treat the subject conceptually and ideally without considering the reality of what may be attainable in a given organization. Consider these two real-world situations.

In a major company with exceptionally good OSHA recordable and lost workday case rates, the chief executive officer declared that results were still not acceptable and that significant reductions in injuries and illnesses were to be made. The extensive and well-qualified staff of safety professionals convinced management to use incident investigation as one means of reinforcing its intent to achieve better results.

A new procedure was put in place to have all OSHA recordable incidents, and those incidents which were judged to have severity potential but did not result in harm or damage, investigated by a team, the talent for which would be selected in relation to the specifics of the incident. For other incidents, the practice of having immediate supervisors make investigations was continued. Re-training sessions on incident investigation were conducted.

A safety professional has to sign off on each completed report, as does the location manager. For a lost workday case, the location manager is required to report personally to the executive higher up in the reporting structure.

A high level of sophistication in incident investigation is attained in this organization. A similar level of sophistication exists only in a small percentage of organizations.

Another entity has 8,000 employees at forty locations, none of which has more than 290 employees. One safety professional is employed—at headquarters. Assume that senior management, because of higher workers compensation costs, says that location managers are to be held to a higher level of accountability for incident experience. Incident investigation guidelines are to be rewritten and the safety professional is to develop a training program and lead training sessions in the six largest locations. Training is to be given at the other locations by resident human resources people.

Surely the level of sophistication attainable in this second situation will be much lower. That should be taken into account by the safety professional in the process of developing and implementing the incident investigation guidelines and the training program. It is also important in that process to try to assess what changes were to take place, actually, in the accountability system. What the boss wants done will be done. If the boss doesn't hold people accountable for top-quality incident investigation, the results will inevitably be less than desirable.

Being Sympathetic toward Supervisors

Another major reality that must be considered is that supervisors may complete only one, two, or three incident investigation reports a year. Although training may be given, knowledge retention is a problem. A properly drafted investigation report form and a causal factor and corrective action refresher guide would be of great help to supervisors who make investigations infrequently.

As the actuality of incident investigation practices is considered, one can't help but feel sorry for the supervisor. It is commonly said in the literature, as Ted Ferry did in *Modern Accident Investigation and Analysis: An Executive Guide* (1), that the supervisor is closest to the action, that the mishap takes place in the supervisor's domain, and that initial responsibility for investigation is very often assigned to the supervisor. Ferry was close in his estimate that first-line supervisors investigate about 90 percent of reported incidents.

Ferry also wrote that if it is the supervisor's duty to investigate, then the supervisor has every right to expect management to provide the training required for the task. A minimum goal should be to provide those who have responsibility to investigate incidents with sufficient knowledge and tools to be effective when making the simpler investigations.

Adopting from What Has Been Learned

As a safety professional attempts to improve the quality of incident investigation, consideration should be given to what has been learned about incident causation in recent years. A good example of the lessons learned derives from ergonomics, which has emerged as a more significant segment of the practice of safety.

Ergonomics is design-based. A proper dealing with ergonomics situations promotes an appropriate balance between the design and engineering, operations management, and task performance aspects of causation.

For instance, incident investigation methods should direct inquiry very early on into what may have been "programmed" into work systems through the design of the workplace or work methods. They should promote inquiry that determines whether the design of the work is error conducive. For all injuries, the first question should be, Are there workplace design or work methods design implications?

To extend that idea, this example is quoted from an essay by Alphonse Chapanis titled "The Error-Provocative Situation: A Central Measurement Problem in Human Factors Engineering." It is from *The Measurement of Safety Performance* (2) by Dr. William E. Tarrants.

> . . . six infants had died in the maternity ward . . . because they had been fed formulas prepared with salt instead of sugar. The error was traced to a practical nurse who had inadvertently filled a sugar container with salt from one of two identical, shiny, 20-gallon containers standing side by side, under a low shelf in dim light, in the hospital's main kitchen. A small paper tag pasted to the lid of one container bore the word "Sugar" in plain handwriting. The tag on the other lid was torn, but one could make out the letters "S...lt" on the fragments that remained. As one hospital board member put it, "Maybe that girl did mistake salt for sugar, but if so, we set her up for it just as surely as if we'd set a trap."

For how many hazards-related incidents has the work situation "set a trap?" In Chapter 8, "Incident Investigation: Studies of Quality," I wrote that in a substantial number of the investigation reports collected, entries suggested that further inquiry should have been made into the design of the workplace or work methods. In many cases, those incidents occurred in error-provocative situations. Yet, in most of those instances, only unsafe acts of employees would be recorded as the incident causes.

Dr. Chapanis (2) offered four axioms that deserve thought both in determining what incident causation model is to apply and in drafting an incident investigation report form.

> Axiom 1. Accidents are multiply determined. Any particular accident can be characterized by the combined existence of a number of events or coincidence of a number of events or circumstances.
>
> Axiom 2. Given a population of human beings with known characteristics, it is possible to design tools, appliances, and equipment that best match their capacities, limitations, weaknesses.
>
> Axiom 3. The improvements in system performance that can be realized from the redesign of equipment are usually greater than the gains that can be realized from the selection and training of personnel.
>
> Axiom 4. For purposes of man-machine design there is no essential difference between an error and an accident. The important thing is that both an error and an accident identify a troublesome situation.

Resources on Incident Investigation

Very few safety texts treat incident investigation in depth. For those who would study the subject, these references are recommended.

- *Modern Accident Investigation and Analysis: An Executive Guide* (1) by Ted S. Ferry

  This text helps in thinking about how incidents occur and how they should be investigated. These are two excerpts.

> Accident prevention depends to a large degree on lessons learned from accident investigation. . . . We cannot argue with the thought that when an operator commits an unsafe act, leading to a mishap, there is an element of human or operator error. We are, however, decades past the place where we stopped there in our search for causes.
>
> While the traditionalist will seek unsafe acts or unsafe conditions, the systems person will look at what went wrong with the system, perceiving something wrong with the system operation or organization that allowed the mishap to take place.

  Ferry wrote extensively about System Safety, Change Analysis, the MORT Process, Multilinear Events Sequencing, and D. A. Weaver's Technic of Operations Review (Weaver, 3).

- *Investigating Accidents with STEP* (4) by Kingsley Hendrick and Ludwig Benner, Jr.

  The heart of the book is its presentation of Sequentially Timed Events Plotting (STEP). The accident investigation methodology presented relies on a new conceptual framework, building on system safety technology and the safety assurance systems of the Management Oversight and Risk Tree (MORT).

- *MORT Safety Assurance Systems* (5) by William G. Johnson

  This text serves well both for incident causation model building and for incident investigation. The accident investigation chapter states that while accident investigation has always been a major element in safety, pre-accident hazard analysis is preferable.

  MORT is an incident investigation and analysis technique that promotes a thorough inquiry into the multiplicity of causal factors and treats, in good balance, design and engineering, and operations management, and task performance practices.

MORT in its entirety is a bit much for the majority of incidents. Nevertheless, MORT offers a sound thought process on which to build an incident investigation system.

- *Guide to Use of the Management Oversight and Risk Tree* (6), a Department of Energy publication

  This is an excellent resource. The MORT system starts with the assumption that an incident has occurred. This user's guide provides explanatory data for a logic diagram through which a review can be made of over 1,500 possible causal factors.

- *Accident Investigation, 2nd edition,* (7) National Safety Council

  This fifty-one page publication is highly recommended as a reference for those who would evaluate and improve their incident investigation systems. A "Guide for Identifying Causal Factors and Corrective Actions" is its centerpiece. Use of the guide requires a systematic approach to identifying multiple causal factors and selecting corrective actions. There are four major sections in the guide:
  - Equipment
  - Environment
  - People
  - Management

For each category, there is a listing of questions under the caption "Causal Factors." Opposite the questions are listings of "Possible Corrective Actions." These two abbreviated examples come from the guide.

Causal Factors	*Possible Corrective Actions*
Did the design of the equipment/tool(s) create operator stress or encourage operator error?	Review human factors engineering principles. Alter equipment/tool(s) to make it more compatible with human capability and limitations. . . . Encourage employees to report potential hazardous conditions created by equipment design.
Were any tasks in the job procedure too difficult to perform (for example, excessive concentration or physical demands)?	Change job design and procedures.

In a form appropriate to the organization to which a safety professional provides counsel, a system of the sort outlined in this National Safety Council publication should result in a focus on actual causal factors and on the development of effective corrective actions. For those who make incident investigations infrequently, a modification of the "Guide for Identifying Causal Factors and Corrective Actions" would serve as a valuable memory jogger. Such a guide is included in this chapter.

Choosing a Causation Model

When designing an incident investigation and analysis process, a determination must be made concerning the causation model on which the process is to be based. What the process designer believes are the facts about incident causation has to be established before an instructional guide can be written or a training program developed. Consider these extremes:

Assume that a decision is made that Heinrich's principles of incident causation are to be the bases for the design of an incident investigation system. Then, the system would stress avoiding "man failure," a Heinrichian term. Emphasis would be on eliminating unsafe acts of employees, through training or some form of behavior modification.

Or, assume that the causation model is to relate to the concepts on which the Management Oversight and Risk Tree (5, 6) is based; the system safety idea; and an understanding that causal factors may derive from workplace or work methods design.

To assist in deciding on the causation model to be applied, I suggest a review of Chapter 6, "Observations on Causation Models for Hazards-Related Incidents," and Chapter 7, "A Systematic Causation Model for Hazards-Related Occupational Incidents."

Making a Self-Audit

I propose that the system designer make a review of the findings in Chapter 8, "Incident Investigation: Studies of Quality," and relate them to the existing incident investigation system. Then I suggest a self-audit of the quality of the incident investigation system in place, using the following self-audit outline.

QUALITY OF INCIDENT INVESTIGATION SELF-AUDIT OUTLINE

1. First, which incident causation model is to apply must be determined. Everything following in this outline regarding causal factors will relate to that model.
2. Take a sample of incident investigation reports completed by supervisors and rate the following for quality:
   a. General information data
   b. Incident descriptions
   c. Recording of injury and illness data

   d. Completion of code entry requirements
   e. Causal factors determinations
   f. Actions to be taken to prevent recurrence
   g. Completion of questions particular to the organization
   h. The quality of the report, overall
3. Using the same sampling of reports, make these assessments:
   a. Are the terms on the forms understood?
   b. Are requirements for causal entries on the forms too simplistic?
   c. Do requirements for causal entries fit with the concepts of incident causation to be applied?
   d. Does the quality of the reports show a need for additional training in incident investigation?
   e. Do the contents of forms indicate a need for additional job hazard analyses?
   f. Does the data gathered allow meaningful analyses?
   g. Is the needed managerial attention given to reports?
   h. Are safety personnel fulfilling their responsibilities?
4. Consider the following questions concerning written incident investigation procedures or guidelines, and the incident investigation report:
   a. Do they clearly convey an understanding of the causation model to be applied?
   b. Are the definitions clear? Is the language ambiguous?
   c. Do they properly balance consideration of design and engineering practices, management and operations practices, and personnel task performance practices?
   d. Do they clearly establish which incidents are to be investigated?
      - Personal injuries (at what level)
      - Property damage (at what level)
      - Environmental incidents (at what level)
      - A frequency of minor personal injuries, property damage, or environmental incidents
      - Incidents not resulting in injury or damage, but which may have severity potential in other circumstances
   e. Are responsibilities for investigation clearly defined?
      - Supervisors
      - Upper management
      - When incident teams are to be used
      - Safety personnel
   f. For report distribution:
      - Are the approval levels what they should be?
   g. For corrective action:
      - Are responsibilities precisely established?
      - Is the follow-up procedure appropriate?
   h. Do the procedures or guidelines and the incident investigation form need revision and re-issuance?

5. Concerning training in incident investigation:
   a. Should the program be improved?
   b. Has it been given to all who need training?
   c. Are refresher courses given?
   d. Is an additional focus on training needed?
6. Is there a need to convince management that giving greater significance to quality incident investigation will assist in their achieving their goals?
7. Having made this self-audit, what actions are to be taken to improve the system?

In determining the actions to be taken, a review of what follows in this essay could be of assistance.

Establishing the Purposes of Incident Investigation

In the guidelines to be published and in training programs, the purposes and objectives of incident investigation should be established and communicated in a language understood in the organization, particularly in relation to its established safety policies and procedures. Some such statements of purpose, adapted from several incident investigation procedures, follow. Choose and adopt from those that are helpful:

- One fundamental element of our comprehensive environment, safety, and health endeavors is the investigation of incidents that may result in harm or damage to people, property, or the environment.
- The objective of this (guideline, standard, procedure) is to establish a uniform method for incident investigation that:

  Reduces injuries and illnesses
  Identifies hazards in the workplace
  Implements consistent reporting
  Permits evaluation of the preventive actions taken
  Provides lessons learned that can be applied to other operations
- A systematic approach to incident investigation, a proper identification of causal factors, and a follow-through on the implementation of corrective actions is essential to a good safety engineering and management system.

  These incident investigation concepts and procedures are to:
  Prevent the human suffering that results from occupational incidents
  Identify causal factors (basic, primary, immediate, secondary, contributing, ancillary)
  Provide for the identification of corrective actions
  Establish an accountability system for proper treatment of the corrective actions proposed
  Prevent similar incidents from occurring

Provide the information necessary for report preparation and record-keeping
Allow the accumulation of data for subsequent analysis

- The primary purpose of incident investigation is to prevent similar occurrences and to discover hazards. Collateral purposes, deriving from a study of the nature of the incident and from identifying causal factors, are to improve policies and standards and their implementation. Additional benefits include impressing on employees the depth of management concerns and improving general performance, supervision and management abilities.

Defining the Terms to Be Used

Keeping in mind the supervisors who complete just one, two, or three incident investigation reports a year, and for whom the training given may be less than the safety professional desires, attaining an understanding of the terms used in procedures and report forms can be a problem.

Whatever the terms used and their complexity or simplicity, their descriptions should be ultimately clear to assure that they communicate what is intended. After reviewing over 900 incident investigation reports, I suggest we not assume that we communicate effectively through the terms we use.

Informing on What Incidents Are to Be Investigated

Texts on incident investigation often propose the improbable when informing on which incidents are to be investigated. These are typical statements, as they appear in the texts: all accidents are to be investigated, regardless of the injury or property damage; and, the policy should be to investigate all accidents.

It is impractical and uneconomical for an entity to try to achieve the theoretical ideal and investigate every incident that results in even the slightest property or environmental damage. Resources are always limited, and priority setting is a must. Some things that are desirable won't get done; choices have to be made. A composite follows of designations actually used by safety professionals when giving instructions on the categories of incidents to be investigated. For each category, definitions would be given.

- First aid treatment
- Minor injuries or illnesses requiring treatment by a doctor
- A frequency of minor injuries or illnesses
- Injury resulting in restricted activity or job reassignment
- Any injury that prevents return to work

- Injury requiring hospitalization
- OSHA recordable incident
- OSHA reportable incident
- An incident resulting in injury to two or more employees
- A fatality
- Minor property damage, product loss, or fires (dollar value set)
- Major property damage, product loss, or fires (dollar value set)
- Environmental spills or releases (air, water, land)
- Incidents that did not result in personal injury or property or environmental damage, but could have had serious consequences had circumstances been different

It's suggested that a higher than usual priority be given to those incidents that did not result in injury or damage but had the potential to do so under other circumstances. Resources should be allocated to those incidents which, if repeated, have the potential to result in serious injury or damage. Severity potential is the proper determinant.

Investigation and Reporting Responsibility

Since hazards-related incidents occur during the work activity for which supervisory personnel have initial responsibility, it follows that they, those closest to operations, should have initial responsibility for incident investigation. (Of course, it would be established in an organization that employees are to report all incidents to their supervisors.)

Guidelines on incident investigation should categorize the incidents for which reporting is only to be by supervisors to location management, and those for which location management is to report upward in the organizational structure, and how soon. Also, incident categories requiring that an investigation team be gathered should be defined.

In a few organizations, the term *major incident,* or something similar, is used to designate the types of incidents for which immediate upward reporting is necessary. This is a composite of the categories of major incidents that safety professionals have set forth in their guidelines.

Major Incident

- An OSHA reportable incident
- Hospitalization of an employee for more than three days
- An incident resulting in injury to three or more employees
- A fatality

- An incident, including a fire, resulting in property damage in excess of $10,000
- A product loss valued in excess of $10,000
- An environmental incident (air, water, land) that must be reported to a governmental authority
- An incident that required building or job site evacuation
- An incident that required emergency shutdown of operations
- An incident that could incite public interest
- An extraordinary or unusual incident creating a crisis or significant emergency
- An incident that did not result in harm or damage but could have had serious consequences under other circumstances
- An incident that will provide a lesson learned for other locations

Incident Investigation Teams

Incident categories requiring investigation by teams should be precisely defined. Such teams may include employees directly involved, supervisory personnel, technical specialists, safety professionals, union personnel, middle management, and possibly the plant manager and headquarters personnel.

A team chairperson with management status, and having the necessary authority and experience for the task, should be designated and given the responsibility for assembling the team. The team chair should:

- See that an appropriate action plan, with specifics suitable to the situation, is developed and distributed, to include:
  - Identification of team participants
  - Meeting schedule
  - Arrangements for meeting facilities
  - Transportation, suggested or arranged
  - Clearance for team members to visit the incident site
  - Arrangement of photographic needs
  - Liaison with the press and public agencies
- Arrange liaison with employee representative(s), government agencies, and news media
- Call and preside over meetings
- Control the scope of team activities by identifying the line of investigation to be pursued
- Assign tasks and establish a schedule
- Assure that no potentially useful data source is overlooked
- Keep interested parties advised of progress
- Oversee the preparation of the final report

Qualifications to be considered when selecting the team members should include:

- Relevant knowledge and experience
- Familiarity with design and engineering, the work methods, the process or operation
- Impartiality
- Analytical capability
- Ability to work with others

Immediate Actions to Be Taken

An understanding is necessary of the actions to be taken by the supervisor immediately responsible, and other management personnel immediately after knowledge is received that an incident has occurred. First priority is to properly care for the safety of employees and members of the public. In outlining immediate actions that supervisors and others are to take, consideration would be given to including these points:

- Immediately visit the incident location
- Determine whether anyone was injured
- If necessary, undertake rescue operations
- Arrange for medical attention if needed
- Determine whether the area should be secured, barricaded, or otherwise isolated
- Arrange for collection and preservation of material relative to the incident
- Prevent use of any equipment involved
- Shut down the operation, if necessary

Notification Procedures

For those incidents for which upward reporting is necessary, a notification procedure, *kept current,* should be in place to which all employees and location management personnel can refer, to assure that all interested personnel receive proper notification. Such a notification procedure would:

- Identify who is responsible for the reporting
- Include the names and phone numbers of the people to be notified (senior executives, corporate environmental, health, and safety personnel, legal department, public relations, union personnel)
- State the communication means to be used for reporting
- Give instructions on how the reporting is to be accomplished during other than usual business times (evenings, weekends, holidays)

- Outline what is to be reported: date and time the incident occurred; location; description of the incident; identification data for employees injured (name, badge number, regular occupation, length of employment, length of time on the job involved); estimated potential injury, property damage, or environmental damage; immediate actions taken; further short-term actions to be taken; reports made or to be made to OSHA, EPA, insurance carriers; actions taken to assemble an investigation team

Fact Finding—Avoiding the Impression of Placing Blame

An incident investigation is made to identify the reality of the causal factors and to recommend corrective actions that will reduce the probability of similar incidents occurring. Under no circumstances should investigations be viewed as having the purpose of placing blame. And this principle has to be imparted through incident investigation guidelines and training programs to all who are to make incident investigations.

Incident investigation requires a particular skill to assure that injured persons and those involved with the incident will be forthcoming with the pertinent causal factors. If an impression is created that the intent is to place blame, persons involved will usually become protective of their well-being, important information may be withheld, and the probability of having a valid investigation will be diminished.

Incident Investigation Techniques

Instruction should be given on how to go about doing the investigation job effectively. The following guidelines apply, after following the procedures set forth under "Immediate Action to Be Taken":

- Timeliness is important—the work should be started before the scene has changed and memories are dimmed.
- Identify your purpose—you are seeking information to identify causal factors to prevent recurrence, not to place blame or find fault.
- Describe the incident, from beginning to end.
- Seek the versions of those involved, those who are to be interviewed; don't try to influence what they say.
- Ask open-ended questions: what, why, when, where, how.
- Confirm your understanding of what has been said, repeating back what you thought you heard.
- Take great care with physical evidence.

- Assume that those interviewed have knowledge of value, and seek from them suggestions on the possible causal factors, to include all elements that contributed to the incident process: the characteristics of things (hardware, equipment, materials), and management, operational, and personnel task performance practices.
- Again, assume that those interviewed have knowledge of value and seek suggestions on the corrective actions that should be taken.
- Above all, be tactful, don't blame or threaten.
- Take good notes: do not use a recorder unless it was agreed that you could do so.
- Review and validate the findings.
- Initiate taking the corrective actions identified .
- Complete the incident investigation report.

Incidents Usually Have Multiple Causal Factors

Seldom do hazards-related incidents have only one causal factor. Incident causation can be complex, even for incidents that do not seem complex. Yet there is an expressed desire in some organizations for simplicity in causal factor determination. In procedure guides, the instruction is given to "identify the cause"—the phrase being singular.

If the single causal factor identified is an employee "unsafe act," the investigation process, although not intended to, may seem to place blame because it concentrates on the behavior of an individual, rather than on the work system.

If people who make incident investigations are encouraged to take the easy route through the instructions given in incident investigation forms and through the investigation procedures established, and to select but one superficial causal factor, more than likely that is what they will do. And that diminishes the value of the investigation. Instructions given in procedure guides should clearly establish that:

- almost all incidents will have multiple causal factors;
- causal factors may include acts of omission or commission concerning design and engineering, and operations management, and task performance practices.

Designer Incident Investigation Reports, and a Reference for Causal Factors and Corrective Actions

What follows is intended particularly for the very large majority of entities that have not or cannot achieve the performance level expected if all investigations were made by safety professionals or trained investigation teams.

Thought was given in developing this material to what realistically can be expected of supervisors, middle managers, and location managers in such organizations; and the importance of providing them with supportive information that leads to more effective causal factor and corrective action determination.

Also, it was recognized that an increasing number of safety professionals now have responsibilities including environmental affairs as well as safety and health, and perhaps fire prevention and protection. Thus, they have a need to provide guidance on reporting for a broad range of incident types.

Since personnel who make very few incident investigations in a year would benefit from having a thought prompter, a draft of a Reference for Causal Factors and Corrective Actions is provided.

Subjects to be included in an incident investigation report should be contained within the front and back of one page, leaving room for comment. For some items in the following list, explanatory comment is given typical to that found in procedure guidelines on incident investigation. It is understood that other items could be added to this list to meet organizational needs.

Incident Investigation Report Subjects from Which Selections Can Be Made

1. Name, title, and department of person completing the report, and the date the report is prepared.
2. Facility, division, department: enter the location at which the incident occurred, the division within the company, and the department within the location.
3. Name of injured person: enter employee's full name last name, first name, and middle initial.
4. Social security number: personnel will have it, if unknown.
5. Age: enter number of years (e.g., 40), not birthday.
6. Sex.
7. Clock number.
8. Employee category: full-time, part-time, temporary, hourly, salaried.
9. Date incident occurred: enter the exact date of occurrence.
10. Shift and time the incident occurred: enter shift number (1, 2, 3) and the time (A.M., P.M.), as closely as possible.
11. Date incident reported: this is the date the incident is first known to the person completing the report.
12. Exact location of the incident: enter precise location (i.e., at weighing station 4).
13. Type of incident: enter injury or illness, fire, property damage, environmental incident, product loss, or significant event that did not result in harm or damage but had the potential to do so.

14. Severity of injury, damage, or release: be as precise as possible; list all personal injuries or illnesses, and give an estimate of the value of the property damage, or details on the amount of the spill or release.
15. Date lost time (disability) began.
16. Date released to regular work or selected work.
17. Was incident reported to a government entity: yes or no. If yes, comment fully in narrative.
18. Employee's usual job title: to what occupation is employee usually assigned?
19. Job title at time of incident: to what occupation was employee assigned when incident occurred?
20. Time in job held when incident occurred: how long had the injured person been assigned to that job?
21. Length of company service: give the number of years the injured person was employed by the company.
22. Is there a JHA for this job?: yes or no. If no, and the company requires job hazard analyses, further comment would be expected.
23. Is there a written procedure for this job?: yes or no. If no and the company requires written job procedures, further comment would be expected.
24. Describe the incident in detail: record everything relevant that preceded the incident; what the persons involved did or failed to do; and what equipment, materials, or aspects of the work environment were involved.

(At this point, the designer of the form has at least two options. One is to give instructions such as those in items 25 and 26, and provide a separate, detailed Reference for Causal Factors and Corrective Actions. Or, the form designer could select from the Reference for Causal Factors and Corrective Actions those subjects appropriate to operations and enter them directly on the form, preceded by check-off boxes, as a part of items 25 and 26. A form can be designed, albeit using small print, that includes abbreviations of twenty or more of the causal factors and all of the corrective actions included in the exhibit that follows, leaves enough room for narrative comment for all but unusual incidents, and still fills just the front and back of a page.)

25. Causal factors: enter *all* of the causal factors (circumstances, events, design of tools, equipment, or the work area, condition of equipment, properties of materials, work methods considerations, management systems, actions or inactions of people) that contributed to the incident. *Almost always, there will be multiple causal factors.*
26. Corrective actions: a) record actions already taken; b) list actions to be taken; c) identify persons responsible for items in b); and d) give expected completion dates. *Typically, several corrective actions are identified through the investigation process.*
27. Investigation report approvals: give names, titles, and approval dates. The approval structure would be set forth in issued procedure guidelines.

If a Reference for Causal Factors and Corrective Actions as thorough as the following is to be provided, it could be included as pages three and four of the Incident Investigation Report Form. (In readable print, it can easily be contained in two pages.) Or, it could be a stand-alone reference (perhaps encased in plastic for preservation purposes) to be maintained in a suitable file, or just included in the procedure manual containing the instructions on incident investigation.

I emphasize that this Reference for Causal Factors and Corrective Actions is presented as a resource for the designer of an incident investigation system, with the assumption that the designer would make revisions in it to suit organizational needs. Whatever option is selected, provision would be made for the person completing the investigation report to enter check marks indicating that each subject had been considered, and also identifying those that were applicable.

Reference for Causal Factors and Corrective Actions

Workplace Design Considerations

1. Hazards derive from basic design of facilities, hardware, equipment, or tooling.
2. Hazardous materials need attention.
3. Layout or position of hardware or equipment presents hazards.
4. Environmental factors (heat, cold, noise, lighting, vibration, ventilation, et cetera) present hazards.
5. Work space for operation, maintenance, or storage is insufficient.
6. Accessibility for maintenance work is hazardous.

Work Methods Considerations

1. Work methods are overly stressful.
2. Work methods are error-inducive.
3. Job is overly difficult, unpleasant, or dangerous.
4. Job requires performance beyond what an operator can deliver.
5. Job induces fatigue.
6. Immediate work situation encouraged riskier actions than the prescribed work methods.
7. Work flow is hazardous.
8. Positioning of employees in relation to equipment and materials was hazardous.

Job Procedure Particulars

1. No written or known job procedure.
2. Job procedures existed but did not address the hazards.

3. Job procedures existed but employees did not know of them.
4. Employee knew job procedures but deviated from them.
5. Deviation from job procedure not observed by supervision.
6. Employee did not match work procedures or equipment.
7. Employee not capable of doing this job (physical, work habit, or behavioral reason).
8. Correct equipment, tools, or materials were not used.
9. Proper equipment, tools, or materials were not available.
10. Employee did not know where to obtain proper equipment, tools, or materials.
11. Employee used substitute equipment, tools, or materials.
12. Defective or worn-out tools were used.

Hazardous Conditions

1. Hazardous condition had not been recognized.
2. Hazardous condition was recognized but not reported.
3. Hazards condition was reported but not corrected.
4. Hazardous condition was recognized but employees were not informed of the appropriate interim job procedure.

Personal Protective Equipment

1. Proper personal protective equipment (PPE) not specified for job.
2. PPE specified for job but not available.
3. PPE specified for job, but employees did not know requirements.
4. PPE specified for job, but employees did not know how to use or maintain.
5. PPE not used properly.
6. PPE inadequate.

Management and Supervisory Aspects

1. General inspection program is inadequate.
2. Inspection procedure did not detect the hazards.
3. Training as respects identified hazards not provided, inadequate, or didn't take.
4. Maintenance with respect to identified hazards is inadequate.
5. Review not made of hazards and right methods before commencing work for job done infrequently.
6. This job requires a job hazard/task/ergonomics analysis.
7. Supervisory responsibility and accountability not defined or understood.
8. Supervisors not adequately trained for assigned safety responsibility.
9. Emergency equipment not specified, not readily available, not used, or did not function properly.

Corrective Actions to Be Considered

1. Job study to be recommended: job hazard/task/ergonomics analysis needed.
2. Work methods to be revised to make them more compatible with worker capabilities and limitations.
3. Job procedures to be changed to reduce risk.
4. Changes are to be proposed in work space, equipment location, or work flow.
5. Improvement is to be recommended for environmental conditions.
6. Proper tools to be provided along with information on obtaining them and their use.
7. Instruction to be given on the hazards of using improper or defective tools.
8. Job procedure to be written or amended.
9. Additional training to be given concerning hazard avoidance on this job.
10. Necessary employee counseling will be provided.
11. Disciplinary actions deemed necessary, and will be taken.
12. Action is to be recommended with respect to employee who cannot become suited to the work.
13. For infrequently performed jobs, it is to be reinforced that there is to be a pre-job review of hazards and procedures.
14. Particular physical hazards discovered will be eliminated.
15. Improvement in inspection procedures to be initiated or proposed.
16. Maintenance inadequacies are to be addressed.
17. Personal protective equipment shortcomings to be corrected.

Training in Incident Investigation

It is fundamental that those who are to investigate incidents have the training and retraining and the practical experience necessary to acquire the special knowledge and skills needed. I recommend that training programs be tailored precisely to the actual needs of those being trained and that the organization's incident experience and a study of the hazards in its operations be considered in the tailoring. I admit that what I suggest will be unattainable, sometimes.

Having reviewed four purchasable incident investigation training programs, all with video cassettes and leader and trainee workbooks, the best that I could say is that they're all right. For each of them, I thought that I would want considerable supporting discussions of incident data that pertained to the actual experience in the entity for which the training was being given, and of its hazards and risk problems.

As a generality, the training programs I reviewed were superficial for causal factor determination. A training program, to be of value, must provide knowledge of incident causation and what the corrective actions ought to be.

It is easy to write that if incident investigation is to be done well, effective training must be provided. Where training resources are limited, the choice in crafting a program should be for that which is attainable. Still, Ted Ferry was right when he wrote: if it is the supervisor's duty to investigate, then the supervisor has every right to expect management to provide the training required for the task (1).

Fortunately, it is not the norm for hazards-related incidents to occur frequently in a given operation. But, that presents another problem. Knowledge acquired through a training program may not be retained by those who make investigations infrequently. Thus, retraining is necessary, and a reference source should be available that helps on the few occasions when investigations are made.

CONCLUSION

Incident investigation, done well, gives positive messages within an organization concerning management's intent for safety. This essay is designed to assist in having it done better.

REFERENCES

1. Ferry, Ted S. *Modern Accident Investigation and Analysis: An Executive Guide.* New York: John Wiley & Sons, 1981.
2. Chapanas, Alphonse. "The Error-Provocative Situation: A Central Measurement Problem in Human Factors Engineering." In *The Measurement of Safety Performance,* William E. Tarrants, ed. New York: Garland Publishing, 1980.
3. "Technic of Operations Review—TOR Analysis." D. A. Weaver. *Journal of the American Society of Safety Engineers* (June).
4. Hendrick, Kingsley, and Ludwig Benner, Jr. *Investigating Accidents with STEP.* New York: Marcel Dekker, 1987.
5. Johnson, William G. *MORT Safety Assurance Systems.* New York: Marcel Dekker, 1980.
6. Guide to Use of the Management Oversight and Risk Tree. SSDC-103. U.S. Department of Energy, Office of Safety and Quality Assurance, Idaho National Engineering Laboratory, Idaho Falls, Idaho, 1994.
7. *Accident Investigation. 2nd ed.* National Safety Council. Itasca, IL, 1995.

Chapter 10

Applied Ergonomics: Significance and Opportunity

Introduction

As applied ergonomics became a major element in the practice of safety in recent years, opportunities arose for achievement, recognition, and professional satisfaction far beyond what has been typical for safety professionals.

Successes from many ergonomics applications that were initiated to resolve injury and illness problems have also achieved increases in productivity and reductions in costs. That gets management's attention. Safety professionals who attain such successes stress the possible economic gains from ergonomics applications, far more than has been done traditionally. Also, there is a greater awareness now that ergonomics is fundamentally design based, and that effective ergonomics applications require a focus on workplace and work methods design.

Significance of Ergonomics within Workplace Safety

What is the significance of ergonomics-related incidents within the universe of workplace injuries and illnesses? Several members of the OSHA staff say in their speeches that about 50 percent of OSHA reportable cases are ergonomics-related. That ties in closely with the results of a significant study made by a major insurance company (1). Working with a claims cost base of more than 1 billion dollars, it was estimated that about 50 percent of workers compensation claims and 60 percent of total claims costs were ergonomics-related.

Workers compensation costs in the United States for 1995 were about 65 billion dollars. The direct workers compensation costs for ergonomics-related injuries and illnesses, not the total costs, could have been 39 billion dollars.

Discussions were held with several safety directors to determine how they would respond to estimates of 50 percent of workers compensation claims and 60 percent of total claims costs being ergonomics-related. There was general agreement that those estimates were sound, but two cautions were expressed: estimates are applicable if the statistical sample is large enough; and variations by industry could be significant. A professor in an industrial engineering department responsible for courses on ergonomics was asked to comment on these numbers; he said they were close to his assumptions.

Unfortunately, ergonomics is narrowly and inappropriately perceived by some to include only cumulative trauma disorders. Opportunities for risk reduction and improving productivity and cost efficiency are lost if ergonomics concepts are not applied to all aspects of workplace and work methods design presenting excessive biomechanical stresses—cumulative or instantaneous.

As a case in point, back injuries are a debilitating and costly segment of workers compensation claims. Preventing back injuries requires eliminating excessively stressful work by designing jobs to fit the capabilities and limitations of the worker. That's applied ergonomics! *This thinking should be extended to include all incidents involving bodily reaction and overexertion, and all other aspects of work that are overly stressful or error-provocative.*

Assume that on a macro and best-information-available-basis, estimates previously cited are close—that 50 percent of workplace injuries and illnesses and 60 percent of workers compensation costs are ergonomics-related. Then, those who have responsibilities in occupational safety and health must have broad involvement in ergonomics.

Defining Ergonomics

For quite some time, I had defined ergonomics as the art and science of designing the work to fit the worker. Because of lessons learned in the past few years, I now use this definition:

> Ergonomics is the art and science of designing the work to fit the worker—to achieve optimum productivity and cost efficiency, and minimum risk of injury.

In some places, that "art and science" has been called human factors engineering. Dr. Alphonse Chapanis gave this definition of human factors engineering in his article "To Communicate the Human Factors Message, You Have to Know What the Message Is and How to Communicate It" (2).

> Human factors engineering is the application of human factors information to the design of tools, machines, systems, tasks, jobs, and environments for safe, comfortable, and effective human use.

In the same article, Dr. Chapanis expressed these views concerning the relationship between human factors engineering and ergonomics:

> I don't intend to enter into an extended discussion about the differences between human factors and ergonomics. Frankly, I don't think the differences, such as they are, are important. . . . So, though I shall be using the words human factors and human factors engineering in this article, I mean them to apply equally to ergonomics and the practice of ergonomics.

Ergonomics and human factors engineering have become synonymous terms. In this essay, ergonomics will be used exclusively, and it will also mean human factors engineering.

Why Define Ergonomics to Include Productivity and Cost Efficiency?

For the occupational setting, this is an appropriate extrapolation of what Dr. Chapanis has written: the outcome of applied human factors engineering is to make the physical aspects of the workplace and work methods safe, comfortable, and effective for human use.

Ergonomics developed out of the need to make work less stressful and more comfortable. Until recently, finding an emphasis in ergonomics texts on applying ergonomics principles to improve productivity and cost efficiency would be an exception.

Evidence continues to mount in support of dealing with ergonomics problems to improve productivity, cost efficiency, safety, and sometimes quality, as segments of an integrated whole. Indications of that evidence are cited in the following history.

I served as chair of a committee at the National Safety Council whose efforts resulted in the Council creating an Institute For Safety Through Design. In the exploratory work to establish the basis for a proposal to create the Institute, I needed to collect actual cases where initiatives undertaken by safety professionals to resolve injury and illness problems also resulted in improved productivity and cost efficiency.

Eleven safety directors were asked for help. Much to my surprise, every one agreed to send me success stories. That was an important indicator of how their jobs had changed. They were involved in productivity, cost-efficiency, and performance measurement much more than I expected.

Nine of the eleven safety directors actually responded with case studies. (Anyone who has done volunteer work will appreciate that such a response was remarkable.) More surprising was the number of success stories they sent me. I knew that I could use only five or six cases in the proposal being developed. They sent me twenty-two, and offered more if I needed them.

As I relate a few of those cases, keep this in mind: safety professionals initiated the studies because of injury experience; and productivity improved and costs were reduced.

A Computer Manufacturer An electronic panel drilling machine was difficult to operate. A study indicated that workers had to bend and reach to perform tasks, and operator comfort, product quality, and performance suffered. As a result of the study, new machines designed for better body mechanics were acquired. In the first year, labor costs were reduced by $270,000 and yield increased $420,000. Productivity was improved; injuries were reduced.

An Aircraft Manufacturer An assembly operation required standing on a platform to get visual and physical access to parts being assembled. Shoulder strain injuries resulted from installing operations being done above shoulder level. Work methods were redesigned. An assembly stand mounted on a hydraulic height-adjustable cart permitted drilling, rivetting, and installing parts in a hands-free operation performed below shoulder height, with good visual and physical access. No more strain injuries occurred. Production increased from two to four units a week; estimated annual cost saving was $52,800.

A Copier Manufacturer A copier product with highly repetitive assembly tasks was estimated to have workers compensation costs 2.5 times higher than the corporate average. A redesign study revealed how hazards could be reduced *and* how simplifications could be made in product design (reduction in the number of parts, and standardization of parts for multiple use), with a substantial reduction in production costs.

A Construction and Farming Equipment Manufacturer On a backhoe assembly line, the design of the work, done within the operator's cab, required taking stressful and awkward positions and repetitive hand tool use. Back injuries were frequent and none but the youngest employees would accept assignment to the work. Wage rate ergonomic team members and the ergonomics coordinator proposed a new design that permitted the work to be done prior to installation of components in the cab, allowing the assembler to work in a good upright position. A major reduction in production time resulted and hazards were notably reduced.

A Computer Chip Maker Because of the very high cost of constructing chip fabrication facilities, achieving cost-effective design solutions through proactive design activity is essential. Great benefits were achieved when the application of sound engineering criteria resulted in human-compatible designs for manual material handling, work stations, controls and displays, and for maintainability. It was found that the costs of implementing solutions in the early design stages have an additional equipment cost impact of zero to 5 percent, versus the 10 to 20 percent of costs for retrofitting to achieve the same ends.

Several case study contributors were enthused about the improved perception management had of them because their work also influenced productivity and cost efficiency. Emphasizing the possible productivity and cost-efficiency gains from ergonomics applications is a different approach from that found in most of the prominent ergonomics literature. Experience now indicates that it should be the exception when an ergonomics analysis is proposed only for safety purposes.

What is needed is an integrated ergonomics task analysis system that addresses productivity, cost efficiency, safety, and sometimes quality—all in one study. That would be unique.

An Example of Professional Recognition: A Real-World Case

What sort of professional recognition can be attained from successes in applied ergonomics? Consider this real-world situation. A generalist in safety became proficient in applied ergonomics before ergonomics became as fashionable as it is now. To convince management of the efficacy of his ergonomics proposals, his practice for the situations requiring considerable work modification expense was to produce cost/benefit data using only potential injury and illness cost savings in the benefit side of the equation.

The design and engineering staff recognized that many of the ergonomics risk reduction actions taken as a result of the proposals made by the safety professional had also improved productivity and reduced costs. This safety professional was advised by design engineers that, for his ergonomics proposals, he no longer needed to make cost/benefit studies. They would undertake to justify the expense to accomplish what he proposed by determining the expected productivity and cost-efficiency improvements, with injury cost savings data being worked into their computations.

Further, because of his demonstrated capability, that safety professional was asked to participate as a member of the design team working on a new production facility, the understanding being that he was to see that they designed things right.

Applied Ergonomics Considerations for Manual Material Handling

For as long as statistics have been compiled, manual material handling incidents have been prevalent both as to frequency of occurrence and severity of injury within the universe of occupational injuries and illnesses. Many safety professionals now take a different approach to determining causal factors for manual material handling incidents when making incident investigations and analyses, having recognized the importance of applied ergonomics.

As a case in point, reviews made of incident analyses compiled in previous years resulted in the conclusion that, often, when improper lifting by an employee was given as the principal incident causal factor, in reality, the basic causal factor was work methods design. Improper lifting had been programmed into the work procedure.

Consider this example: Bags weighing seventy pounds are at floor level on pallets. Workers slit open the bags, lift them to shoulder height, and pour the contents into hoppers. Speed, stooping, and twisting are required. Back injuries are frequently reported. Causes recorded in investigation reports are always improper lifting by the employee. Invariably, the corrective action is to reinstruct the worker in proper lifting techniques.

For remedial action, another employee training program would be conducted on how to lift safely. The solution to this problem is obviously work methods redesign. No amount of training or reinstruction or behavior modification could be an adequate solution. Yet safety professionals accept that this scenario often relates to the actuality of past investigations, analyses, and corrective actions proposed.

Safety professionals cannot escape the fact that many of the corrective measures proposed in the past to reduce the number of overexertion incidents, particularly back injuries, were not effective. Improving employee selection, training, and behavior are ineffective solutions to work methods design problems. (But it's recognized that, sometimes, they are the only measures that can be taken.)

Negative findings of the National Institute for Occupational Safety and Health (NIOSH) concerning the use of employee selection techniques and training to prevent manual lifting injuries have not been disputed. They are reported in *Work Practices for Manual Lifting* (3). These are some of those findings:

> No significant reduction in low back injuries was found by employers who used medical histories, medical examinations, or low back x-rays in selecting the worker for the job.
>
> And . . . unfortunately, no controlled epidemiological study has validated any of the contemporary theories on (lift techniques).
>
> Also, the importance of training and work experience in reducing hazards is generally accepted in the literature. The lacking ingredient is largely a definition of what the training should be and how this early experience can be given to the naive worker without harm.
>
> More important than proper selection and training in the long-term prevention of accidents and injuries relating to lifting, is providing a safe ergonomics environment in which to work.

There is a need for debate and a learned paper on manual lifting techniques, reflecting NIOSH research and other studies that put past practices in doubt. Valid guidelines for training on manual material handling should be included in the paper.

Participating in the Design Process

I believe that the greatest strides forward as respects safety, health, and the environment will be made through the design processes. Several safety professionals who recently participated in design discussions in their companies, a new venture for them, were euphoric when telling of their experiences. They expressed an added sense of accomplishment because they influenced design engineers to include safety considerations in their decision-making, the result being the avoidance of bringing hazards into the workplace. That's understandable.

Without question, the content of professional safety practice is undergoing a necessary and vital transition. To prepare for the needs and opportunities presented by this evolution, basic ergonomics-related knowledge and skill, as follows, should be acquired by the safety professional who would participate in design decisions.

- Working knowledge of engineering principles is necessary to communicate successfully with engineers who make workplace and work practice design decisions.
- An understanding of the broad range of human physical and psychological variations that employees bring to the workplace (the subject of anthropometry) is necessary.
- Knowledge of the application of mechanical principles for the analysis of forces on body parts will be necessary, and can be obtained through a study of biomechanics.
- Appreciation is needed of the shortcomings of designing a fixed and inflexible workplace to average characteristics—the all-too-typical design practice.
- Through the foregoing, knowledge and skill would be attained concerning what constitutes good workstation and work practice design.

Courses on ergonomics are available in great number. Some are in appropriate depth in relation to needs and opportunities; others are superficial. Considering the impact that ergonomics is having and will have on the content of professional safety practice, the courses of study taken should be substantive.

Acquiring an adequate library of ergonomics-related texts has also become requisite. A recommended list follows, moving from the basic and necessary to the more complex, the last being a design reference.

Recommended Reading

- *Work Practices Guide for Manual Lifting* (3), and *Applications Manual for the Revised NIOSH Lifting Equation* (4), National Institute for Occupational Safety and Health

  A safety professional interested in back injury prevention must be familiar with these NIOSH publications.

- *Cumulative trauma disorders: A manual for musculoskeletal diseases of the upper limbs* (5), edited by Vern Putz-Anderson

  Putz-Anderson's book is an excellent primer and is a must as a reference. Contributors have been prominent in the development of ergonomics concepts. It is well written, and its approach to prevention is sound.

- *Ergonomics: A Practical Guide* (6), National Safety Council

  The National Safety Council's ergonomics guide provides a basic outline for an ergonomics program.

- *Ergonomic Design for People at Work, Volumes 1 and 2* (7), by the Human Factors Section at Eastman Kodak

  Both of these volumes, for which Suzanne H. Rodgers was the technical editor, are practical and easily followed.

- *Fitting the Task to the Man: An Ergonomic Approach* (8) by E. Grandjean

  Considering the number of times Grandjean is quoted in other texts and articles, this text has obviously proven its value as a resource.

- *The Ergonomics Edge—Improving Safety, Quality, and Productivity* (9) by Dan MacLeod

  I quote from the introduction to MacLeod's book. "The goal of this book is to convince you that every business strategy should include improving the user-friendliness of both the workplace and the end product."

- *The Practice and Management of Industrial Ergonomics* (10) by David C. Alexander

- *Fundamentals of Industrial Ergonomics* (11) by B. Mustafa Pulat

- *Human Factors Design Handbook* (12) by W. E. Woodson

  The Practice and Management of Industrial Ergonomics and *Fundamentals of Industrial Ergonomics* are used in ergonomics courses in engineering curricula. Woodson's book is design handbook.

For all of these texts, a common theme is apparent—design the job to fit the worker's capabilities and limitations.

Undertaking an Ergonomics Initiative

What is the preferred order of priority when instituting an ergonomics initiative? To begin with, the data-gathering necessary to identify jobs, operations, and departments that present workstation or work practice problems would be done through:

reviews of incident investigation reports and OSHA 200 Log
reviews of workers compensation and group insurance claims for ergonomics-related incidents
discussion with the personnel staff concerning high turnover jobs and jobs with excessive absenteeism
seeking comments from supervisors on stressful jobs
encouraging employees to identify jobs for which they have experienced work-related pain or discomfort
walk-through tours of the facility to locate jobs that:

- require a great deal of strength or power
- require considerable stretching, bending, or stooping
- require considerable lifting
- require the worker to assume awkward positions
- are extremely repetitive
- may present excessive vibration exposure from power tool usage
- are performed at a rapid pace
- present environmental discomfort (temperature, contaminants, lighting, et cetera)
- prompt employee-generated work changes (benches, wraps on tool handles, cheater bars used on wrenches and valves, padding on chairs, footrests)
- are monotonous
- involve multiples of the preceding considerations

After making such a study, hazards would be analyzed and priorities established for the actions to be taken. In setting priorities, consideration would be given to the most stressful jobs; the probability that severe injuries may result from doing particular jobs; workstations that can be easily modified; and workstation changes that require capital expenditures.

In selecting individual tasks for treatment, these excerpts from *Cumulative Trauma Disorders: A Manual for Musculoskeletal Diseases of the Upper Limbs* (5) give good guidance:

> There is seldom a simple, single change to be made. More often there are numerous overlapping problems involving some combination of high production demand, faulty work methods, awkward work station layouts or ill-fitting tools.

Perhaps the most cautious way to proceed is to administer an ergonomic intervention with the same degree of care as one would use with any new remedy. One should:

(1) Perform a thorough examination (job analysis) first to determine the specific problem;
(2) Evaluate and select the most appropriate intervention(s) (the assistance of an expert may be useful here);
(3) Start conservative treatment (implement the intervention), on a limited scale if possible;
(4) Monitor progress; and
(5) Continue to adjust or refine the scope of the intervention as needed.

After analyzing the data gathered, a communication to management proposing an ergonomics initiative would be prepared. That communication should include a plan that addresses the following elements and how they would be incorporated within the overall safety processes in place:

Management commitment and involvement
Responsibility and accountability
Engineering involvement
Purchasing liaison
Employee involvement
Work site analysis
Engineering modifications
Administrative controls
Personal protective equipment
Training
- Engineers—first training priority
- Supervisors
- Employees

Medical management
Documentation, monitoring, feedback
- Engineering revisions
- Work practice revisions
- Enforcement of safe work practices
- Record keeping, analysis, and review

Safety professionals should recognize that a significant change in direction is being proposed to management when a new ergonomics initiative is suggested. Knowing how to achieve change successfully would serve the entity well in this endeavor.

When introducing an ergonomics initiative, it is important that the infrastructure is in place to respond properly to the work orders that surely will be generated by an increased awareness of ergonomics concepts. Assume that the initiative receives management support and, in time, an employee awareness and education program is conducted. If maintenance and engineering personnel are not equipped to act on proposals for workstation and work practice changes made by employees in a reasonable time, employee enthusiasm will surely be dampened and the credibility of the ergonomics initiative will be damaged.

Only a few years ago, the conventional thinking among many safety professionals was that applied ergonomics would not be a prominent aspect of safety practice. Their assumption was that expenditures for ergonomics improvement would be excessive in relation to the benefits that could be obtained. That argument has no substance since history has proven that many work practice improvements can be made without great expense.

An example of a revision in material handling methods that can be made with minimum expense is described in the paper "Dynamic Comparison of the Two-hand Stoop and Assisted One-hand Lift Methods" (13) by Cook, Mann, and Lovested. That paper resulted from a study made at the University of Iowa testing an alternative one-hand lift method developed to address parts picking in a warehouse setting.

Countless success stories are being told by safety professionals about the suggestions for ergonomics workplace improvements made by first-line employees. Many of these suggestions are easily accomplished, inexpensive, and effective. Employees must be actively involved in ergonomics and their participation should be considered a valuable asset.

Learning from recent history, some cautions are offered respecting the introduction of an ergonomics initiative. First, the initial thrust should be to educate and obtain the support of upper management. Having obtained upper management awareness and support, the next course of action should be to train engineering and maintenance personnel in ergonomics fundamentals. That is the first training priority, and should be done before awareness and training programs are run for supervisors and line employees. It is folly to assume that engineering personnel, even though they may be degreed engineers, have basic ergonomics knowledge.

To repeat, the infrastructure should be in place to handle what could be a flood of work orders when supervisors and line employees become participants in the ergonomics initiative.

OSHA and an Ergonomics Standard

OSHA personnel say they remain committed to the issuance of an ergonomics standard. If that gets done, the standard will have a major impact on both the content of safety practice and the ergonomics processes to be put in place.

OSHA's target date for the issuance of a proposed ergonomics standard has been moved forward several times. Debates continue on what the standard should contain and opinions vary greatly. Comments follow on some possibilities concerning what may eventually become an OSHA ergonomics standard.

In January of 1989, OSHA published voluntary *Safety and Health Program Management Guidelines* (14) for general industry. While a safety professional might fault those guidelines on detail, they are generally sound. If an adaptation of those guidelines became a standard, an ergonomics supplement could easily be fitted to them. That represents the simplest, yet a very sensible, approach.

OSHA's only official guidelines on ergonomics, the *Ergonomics Program Management Guidelines for Meatpacking Plants* (15), were issued in August 1990. In the introduction, recognition is given to the *Safety and Health Program Management Guidelines* (14) issued in 1989. They are "recommended to all employers as a foundation for their safety and health programs and as a framework for their ergonomics programs."

Note the wording "foundation for their safety and health programs and as a framework for their ergonomics programs." It implies establishing a sound safety and health program and merging ergonomics measures within it. That would be good hazards management practice.

Ergonomics guidelines for meatpacking duplicate much of the language in the safety and health management guidelines. And their structures are alike.

Ergonomics Program Management Guidelines for Meatpacking Plants (15) are generic and little modification would be needed for them to serve the purposes of general industry. The table of contents follows, with only two phrases removed to show their adaptability. (In the actual contents, the phrase "for the Meat Industry" follows "Examples of Engineering Controls" in Item III B. In Item III C, "in Meatpacking Establishments" follows "Medical Management Program."

I. MANAGEMENT COMMITMENT AND EMPLOYEE INVOLVEMENT
   A. Commitment by Top Management
   B. Written Program
   C. Employee Involvement
   D. Regular Program Review and Evaluation
II. PROGRAM ELEMENTS
   A. Worksite Analysis
   B. Hazard Prevention and Control
   C. Medical Management
   D. Training and Education
III. DETAILED GUIDANCE AND EXAMPLES
   A. Recommended Worksite Analysis Program for Ergonomics
   B. Hazard Prevention and Control: Examples of Engineering Controls
   C. Medical Management Program for the Prevention and Treatment of Cumulative Trauma Disorders

Having a general industry ergonomics standard in the style of the guidelines for meatpacking could be a second option.

In the fall of 1990, unpublished and unofficial *Ergonomics Program Management Recommendations for General Industry* were written at OSHA. They are labeled "Internal Draft—Not for Public Release." Including attachments and a bibliography, they fill ninety-four double-spaced pages. About thirty-six of the ninety-four pages are devoted to the specifics of medical management.

Overall, the recommendations are a bit much. They include much detail on ergonomics methodology in the form of a specification standard. Nevertheless, they are thought-provoking. They give an indication of the thinking of some individuals at the time the recommendations were written on what should be included in OSHA's ergonomics standard.

The recommendations duplicate a good part of the *Safety and Health Program Management Guidelines* (14). Excerpts representing about one-eighth of the document are included as an addendum to this essay since they embody the management elements of a good ergonomics initiative.

Ergonomics Program Management Recommendations for General Industry portray a third possibility for an OSHA standard.

In 1995, a proposed OSHA ergonomics standard filling over 400 pages was published. It produced much adverse response.

Meeting the Challenge

It should be the goal of safety professionals to be sought for their counsel by those who make workplace and work practice design decisions. Becoming qualified in ergonomics partially prepares safety professionals for that role. Ergonomics, as it continues to emerge and have great influence on the practice of safety, has compelled many safety professionals to again become students. That's necessary to meet the challenge and to maintain a practice of safety at a professional level.

Addendum

Excerpts from OSHA's Unpublished and Unofficial Ergonomics Program Management Recommendations for General Industry

[Editorial note: Everything except the words in brackets is taken verbatim from the recommendations.]

Introduction

Effective management of worker safety and health includes protection from all work-related hazards whether or not they are regulated by specific government standards.

The Occupational Safety and Health Act of 1970 (OSH Act) clearly states that the general duty of all employers is to provide their employees with a work place free from recognized hazards.

In recent years, there has been a dramatic increase in the occurrence of . . . injuries and illnesses due to ergonomic hazards.

The incidence and severity of [ergonomic-related] work place injuries and illnesses . . . demand that effective programs be implemented to protect workers from these hazards. These [programs] should be a part of the employer's overall safety and health management program.

In January 1989 [OSHA] published voluntary *Safety and Health Program Management Guidelines,* which are recommended to all employers as a foundation for their safety and health programs and as a framework for their ergonomics programs.

In addition, OSHA has developed the following general industry ergonomics program management recommendations . . . . [This program] is divided into two primary sections: a discussion of the importance of management commitment and employee involvement, followed by a recommended program with four major elements—worksite analysis, hazard prevention and control, medical management, and training and education.

The science of ergonomics seeks to adapt the job to the worker by designing tasks within the worker's capabilities and limitations.

Experience has shown that instituting programs in ergonomics has significantly reduced the incidents of [ergonomics-related] disorders and, often, improved productivity.

I. Management Commitment and Employee Involvement

Commitment and involvement are complementary and essential elements of a sound safety and health program. Commitment by management provides the organizational resources and motivating force necessary to deal effectively with ergonomic hazards.

Employee involvement and feedback through clearly established procedures are likewise essential, both to identify existing and potential hazards and to develop and implement effective ways to abate such hazards.

A. Commitment by Top Management

The implementation of an effective ergonomics program includes a commitment by the employer to provide the visible involvement of top management. An effective program should have a team approach with top management leading the team, and including the following:

1. Management involvement demonstrated through . . . the priority placed on eliminating the ergonomic hazards.
2. A policy which places safety and health on the same level of importance as production.
3. Commitment to assign and communicate . . . responsibility.
4. Commitment to provide adequate authority and resources.
5. Commitment to ensure that [all are held] accountable for carrying out [their] responsibilities.

B. Employee Involvement

An employer should provide for and encourage employee involvement in the ergonomics program and in decisions which effect the worker safety and health, including the following:

1. An employee complaint or suggestion procedure which allows workers to bring their concerns to management and provide feedback without fear of reprisal.
2. A procedure which encourages prompt and accurate reporting of potential CTD's [cumulative trauma disorders] (and other ergonomics-related injuries).
3. Safety and health committees which receive information on ergonomic problem areas.
4. Ergonomic teams or monitors with the required skills.

C. Written Program

Effective implementation requires a written program for job safety, health, and ergonomics that is endorsed and advocated by the highest level of management and that outlines the employer's goals and plans. The written program should be communicated to all personnel.

D. Regular Program Review and Evaluation

Procedures and mechanisms should be developed to evaluate the implementation of the ergonomics program and to monitor progress accomplished.

The results of management's reviews should be a written progress report and program update, which should be shared with all responsible parties and communicated to employees.

II. Program Elements

An effective ergonomics program should include the following four components: worksite analysis, hazard prevention and control, medical management, and training and education.

A. Worksite Analysis

Worksite analysis provides for both the identification of problem jobs and risk factors associated with problem jobs. The first step is to determine what jobs and work stations are the source of the greatest problems. Thus, a systematic analysis program should be initiated by reviewing injury and illness reports.

The second step is to perform a more detailed analysis of those work tasks and positions previously determined to be problem areas for their own specific ergonomic risk factors.

The analysis should be routinely performed by a qualified person.

The analyst[s] should keep in mind the concept of multiple causation [i.e.] the combined effect of several risk factors . . . jobs, operations, or work stations that have multiple risk factors have a higher probability of causing [ergonomics-related] disorders.

B. Hazard Prevention and Control

Ergonomic hazards are prevented primarily by effective design of a job or job site. An employer's program should establish procedures to correct or control ergonomic hazards using appropriate engineering, work practice, and administrative controls, coordinated and supervised by an ergonomist or a similarly qualified person.

Administrative controls reduce an employee's exposure to tasks with ergonomic hazards by schemes such as rotation to less stressful jobs, reduced production demand or quotas, and increased rest breaks.

Engineering controls, where feasible, are the preferred method of control. The primary focus of an ergonomics program is to make the job fit the person, not force the person to fit the demands of the job. This can be accomplished by redesigning the work stations, work methods, work tools, and work requirements to reduce or eliminate excessive exertion, repetitive motion, awkward postures, and other risk factors.

1. Engineering Controls

a. Principles of Work Methods

Work methods [including workstations and tools] should be designed to reduce [ergonomics] exposure.

The first step is to identify the present problems [and a task analysis should follow].

b. Principles of Work Station Design

Work stations should be designed to accommodate the vast majority of the persons who work at a given job. Because workers vary considerably, it is not adequate to design for the average worker.

Work stations should be easily adjustable and designed for each specific task so that they are comfortable . . . and are appropriate for the job performed.

Specific attention should be paid to static loading of muscles, work activity height, reach requirements, force requirements, sharp or hard edges, thermal conductivity of the work surface, proper seating, support for the limbs, work piece orientation, work piece holding, and layout.

c. Tool and Handle Design

Proper attention should be paid to the selection and design of tools and work station layouts to minimize the risk of cumulative trauma disorders [and other ergonomics-related injuries].

d. Back Injury Prevention

While most back disorders result from cumulative trauma or gradual insult to the back over time, some injuries are caused by a sudden excessive load or fall. These disorders are by far the largest single category of all lost-time injuries, and have enormous financial implications.

Historically, back injury prevention has focused primarily on problems of materials handling. Common preventive measures were:

- Training workers how to lift "safely"
- Restricting the weights lifted to some maximum
- Selecting the "strongest" workers for the "heavy work"

Scientific research, industrial studies, and compensation statistics have demonstrated that these approaches have been ineffective in reducing and controlling the problem. It is now recognized that effective back injury prevention requires ongoing effort with long term commitment to:

- Redesigning existing workplaces, jobs and equipment
- Providing training and education for all members of an organization on the causes and means of preventing back injuries as well as proper individual body mechanics

It should be noted that there are a number of specific risk factors that may act alone or in combination to increase the risk of back disorders. . . A list of relevant job and work station considerations—which is by no means all inclusive—follows.

- Workplace and work station layout
- Actions and movements

- Working posture and work positions
- Frequency and duration . . . . of manual handling activities
- Load considerations

2. *Work Practice Controls*

An effective program for hazard prevention and control also includes procedures for safe and proper work which are understood and followed by managers, supervisors, and workers. Key elements of a good work practice program for ergonomics include proper work techniques, employee conditioning, regular monitoring, feedback, maintenance, adjustments, and modifications, and enforcement.

3. *Personal Protective Equipment (PPE)*

Potential ergonomic hazards should be considered when selecting PPE.

4. *Administrative Controls*

A sound overall ergonomics program includes administrative controls that reduce the duration, frequency, and severity of exposure to ergonomic hazards.

a. Examples of administrative controls include the following:

- Reducing the total number of repetitions
- Providing rest pauses to relieve fatigued muscles
- Increasing the number of employees assigned to the task
- Using job rotation, with caution and as a preventive measure, not in response to symptoms of cumulative trauma disorders

b. Effective programs for facility, equipment, and tool maintenance to minimize ergonomic hazards include the following measures:

- A preventive maintenance program for mechanical and power tools and equipment
- Performing maintenance regularly
- Effective housekeeping programs

C. Medical Management

An effective medical management program for cumulative trauma disorders is essential to the success of an employer's total ergonomic program.

The major components of a medical management program for the prevention and treatment of cumulative trauma disorders are:

1. Periodic Workplace Walkthrough
2. Symptoms Survey

3. Identification of Restricted Duty Jobs
4. Health Surveillance
5. Employee Training and Education
6. Accurate Recordkeeping
7. Periodic Program Evaluation

[Editorial Note: In the recommendations from which these excerpts were taken, thirty-six pages of proposals and comments follow the preceding list of the major components of a medical management program.]

D. Training and Education

The last major program element for an effective ergonomic program is training and education . . . to ensure that employees are sufficiently informed about the ergonomics hazards to which they may be exposed and thus are able to participate actively in their own protection.

A training program should include the following individuals:

- All effected employees
- Engineers and maintenance personnel
- Supervisors
- Managers

References

1. Ergonomics Workshop. Aetna Life and Casualty, Hartford, CT, 1990.
2. "To Communicate the Human Factors Message, You Have To Know What the Message Is and How To Communicate It." Alphonse Chapanis. *Human Factors Society Bulletin* (November 1991).
3. *Work Practices Guide for Manual Lifting.* National Institute for Occupational Safety and Health (NIOSH), Cincinnati, 1981.
4. *Applications Manual for the Revised NIOSH Lifting Equation.* NIOSH, Cincinnati, 1994.
5. Putz-Anderson, Vern, ed. *Cumulative Trauma Disorders: A Manual for Musculoskeletal Diseases of the Upper Limbs.* Philadelphia: Taylor & Francis, 1988.
6. *Ergonomics: A Practical Guide.* National Safety Council, Chicago, 1988.
7. Rodgers, Suzanne H., technical editor. *Ergonomic Design for People at Work,* Vols. 1 and 2. New York: Van Nostrand Reinhold, 1986.
8. Grandjean, E. *Fitting the Task to the Man.* Philadelphia: Taylor & Francis, 1988.
9. MacLeod, Dan. *The Ergonomic Edge: Improving Safety, Quality, and Productivity.* New York: Van Nostrand Reinhold, 1995.

10. Alexander, David C. *The Practice and Management of Industrial Ergonomics.* Englewood Cliffs, NJ: Prentice Hall, 1986.
11. Pulat, Mustafa B. *Fundamentals of Industrial Ergonomics.* Englewood Cliffs, NJ: Prentice Hall, 1992.
12. Woodson, Wesley E. *Human Factors Design Handbook.* New York: McGraw-Hill, 1992.
13. "Dynamic Comparison of the Two-hand Stoop and Assisted One-hand Lift Methods," T. M. Cook, S. Mann, and G. E. Lovested. *Safety and Health Magazine,* National Safety Council, February 1991.
14. *Safety and Health Program Management Guidelines.* Occupational Safety and Health Administration. *Federal Register* 54 (3904), January 26, 1989.
15. *Ergonomic Program Management Guidelines for Meatpacking Plants.* OSHA 3123, U.S. Department of Labor, Washington, D.C.: 1990.

Chapter 11

System Safety: The Concept

INTRODUCTION

Although system safety professionals have achieved notable successes, very few generalists in the practice of safety have adopted their concepts and techniques. But there are lessons to be learned from their successes. I believe that generalists in the practice of safety will improve the quality of their performance by acquiring and disseminating knowledge of what system safety is all about. I do not say that those generalists must become specialists in system safety.

In *MORT Safety Assurance Systems* (1), William G. Johnson made these comments (with which I agree) about accomplishments that could not have been achieved without applying system safety concepts:

> The system safety programs used in aerospace, nuclear, and military projects provided a well-ordered guide to some requirements for a superlative effort. Indeed, they are a route to accomplishing things which would otherwise be beyond human reach.

Those accomplishments are a matter of fact, and they are immense. Anyone who has an understanding of the complexity of the hardware, the demands on the personnel, and the attendant risks must marvel at the success of a space shot.

Influences Encouraging Adoption of System Safety Concepts

In 1993, the American Society of Safety Engineers updated and reissued its paper titled *Scope and Functions of the Professional Safety Position* (2). I quote very briefly from that paper, which is an excellent and current position description for the safety professional.

Scope and Functions of the Professional Safety Position

> The major areas relating to the protection of people, property and the environment are:
>
> A. Anticipate, identify and evaluate hazardous conditions and practices.
> B. Develop hazard control designs, methods, procedures and programs.
> C. Implement, administer and advise others on hazard controls and hazard control programs.
> D. Measure, audit and evaluate the effectiveness of hazard controls and hazard control programs.

Changes from the prior issue of this ASSE paper are subtle, but important. In item A, it now says that the safety professional is to "anticipate hazards." And in item B, the word "designs" is an addition.

To be in a position to anticipate hazards, one must be involved in the design processes. To effectively participate in the design processes, the safety professional must be skilled in hazards analysis techniques. Being a participant in the design processes and using hazards analysis techniques are basic in applied system safety.

These changes in ASSE's version of the *Scope and Function of the Professional Safety Position* reveal how the work of some safety professionals has evolved. They are also predictive of the future, and the opportunities the future holds for enterprising safety professionals.

Joe Stephenson got it right in *System Safety 2000* (3) when he wrote:

> Safety is achieved by doing things right the first time, every time. If things are done right the first time, every time, we not only have a safe operation but also an extremely efficient, productive, cost-effective operation.

For new facilities and equipment, and for their subsequent alteration, the time and place to effectively and economically avoid, eliminate, or control hazards is in the design or redesign processes. Participating in the design processes presents opportunities for upstream involvement by safety professionals using system safety concepts.

Also, there has been a renewed recognition that engineering is the preferred course of action to avoid, eliminate, or control hazards. That renewed recognition derives from several sources, among which are the involvement of safety professionals in applied ergonomics; meeting the requirements for hazards analyses and pre-start-up reviews in OSHA's standard for Process Safety Management of Highly Hazardous Chemicals; and quality management.

A development at the National Safety Council in 1995 is another indicator of the view that great societal benefit can be obtained if hazards are addressed in the design processes. An Institute for Safety through Design was created. These are two of its purposes: to have the concepts of designing for safety included in courses taken by engineering students; and to develop course materials on designing for safety for the thousands of engineers now employed.

Defining System Safety

Unfortunately, the term *system safety* does not convey a clear meaning of the practice as it is applied. Published definitions of system safety are of some help in understanding the concept, but they do not communicate clearly. To move this discussion forward, I quote from two texts on system safety, and one magazine article.

Stephenson, in *System Safety 2000* (3), gives this definition of system safety.

> Simply put, system safety is the name given to the effort to make things as safe as is practical by systematically using engineering and management tools to identify, analyze, and control hazards.

In *System Safety Engineering and Management* (4), Harold E. Roland and Brian Moriarty ask, What is System Safety? In response to their own question, they give two meaningful comments and then establish the system safety objective.

> The system safety concept is the application of special technical and managerial skills to the systematic, forward-looking identification and control of hazards throughout the life cycle of a project, program, or activity.
>
> The system safety concept . . . involves a planned, disciplined, systematically organized and *before-the-fact* process characterized as the *identify-analyze-control* method of safety.
>
> System Safety Objective: *A safety objective such that each person will live and work under conditions in which hazards are known and controlled to an acceptable level of potential harm.*

Richard G. Pearson and Mahmoud A. Ayoub gave this view of the systems approach in "Ergonomics Aids Industrial Accident and Injury Control" (5).

> By systems approach, we mean a conscientious, systematic effort to design an effective system, such as a production plant, giving due consideration to

the interaction among man, machine, and environment. From the ergonomics viewpoint, prime consideration is given to human performance and safety considerations. A cardinal principle of ergonomics is that since everything is designed ultimately for man's use or consumption, man's characteristics should be considered from the very beginning of the design cycle.

. . . systems evaluation should be a continuous process during the design, development, installation, operation, and maintenance of an industrial manufacturing system.

Using those definitions as a base, and with some expansions, I present the following for consideration by safety generalists to emphasize what system safety encompasses and to relate system safety to the workplace and work methods.

The System Safety Idea

1. System safety is hazards and design based.
2. Hazard analysis is the most important safety process in that, if that fails, all other processes are likely to be ineffective (Johnson, 1).
3. Hazards are most effectively and economically anticipated, avoided, or controlled in the initial design processes, or in the redesign of existing systems.
4. All hazards-related incidents result from interaction of elements in a system.
5. Applied system safety requires a conscientious, planned, disciplined, and systematic use of special engineering and managerial tools.
6. On an anticipatory and forward-looking basis, hazards are to be known, avoided, eliminated, or controlled so that systems can be practically designed to attain minimum risk.
7. Applying specifically developed hazard analysis and risk assessment techniques is a necessity in system safety applications.
8. System safety concepts promote the establishment of policies and procedures that are to achieve an effective, orderly, and continuous hazards management process for the design, development, installation, and maintenance of all facilities, materials, hardware, equipment, and tooling, and for their eventual disposal.
9. In system design processes, consideration is to be given to the interactions among humans, machines, and the environment, and the capabilities and limitations of people and their penchant for unpredictable behavior.
10. Applied system safety is to attain safety—that state for which the risks are judged to be acceptable.

I am aware that my outline of the System Safety Idea does not fit precisely with any of the several definitions of system safety published. It encompasses most and goes beyond several. I hope that it's of interest to more generalists in the practice of safety than are now applying system safety concepts. I am confident that the application of system safety concepts in the business and industrial setting would result in significant reductions in injuries and illnesses.

System safety is hazards focused, as are all the subsets of the practice of safety, whatever they are called. System safety commences with hazard identification and analysis. Do that poorly, and all that follows is misdirected. Hazard/risk assessment methods used in system safety have been successful. The generalist in safety practice ought to know more about them.

A System Safety Application Outline

Work done by Ernest Levins, who was director of safety at McDonald-Douglas in Santa Monica, California, helps in thinking about system safety and how the concepts can be applied in the workplace setting. What Levins wrote is expressed in uncomplicated and easily understood terms. In an article titled "Search I—Fourth Installment, Locating Hazards Before They Become Accidents" (6), Levins gave his views on an effective hazard analysis scheme, which follows:

Seven Steps to Follow

1. Define the system in space and time—including its objectives, the location of its interfaces with other systems of interest, and the analytical limits of resolution within the system (may vary, depending on the analyst's interests).
2. Specify the identifiable undesired outcomes, states, or conditions within the defined system.
3. Select the key undesired system-characteristic outcomes that serve as the basis for decision—by comparing them with some index of criticality (e.g., negligible, marginal, critical, catastrophic); record the results.
4. Determine the possible modes of occurrence of the selected undesired outcomes.
5. Evaluate the likelihood of occurrence of the possible modes of occurrence. This can be done with logic alone; failure rate data are not needed in most industrial systems. An estimate of the mission-success can even be made on the basis of events being "likely" or "unlikely" to occur.
6. From the foregoing, decide if the system design is adequate to prevent failure; and, if not, what design changes are required to improve relative safety of the system.
7. Analyze any system revision as above, and repeat as often as necessary until the optimum design is achieved.

Levins speaks of "relative safety" and "optimum design." What do those terms imply? Design goals are not to attain a risk-free system, which is unattainable. In arriving at the optimum design, judgments will be made concerning the probability of hazard realization, the severity of harm or damage that could result, the costs of risk reduction, performance requirements, and scheduling.

Levins does suggest that some of the judgments necessary can be made with logic alone. That's important. System safety concepts can be implemented in many cases without applying extensive and time-consuming hazard analysis and risk assessment methods. But logic alone will not be adequate for all hazards. For some hazards, applying the hazard analysis and risk assessment methods specifically developed in the evolution of system safety will be necessary.

How System Safety Evolved

A bit of history on the evolution of system safety will give an insight into its origins, of the need for the hazard analysis and risk assessment techniques which were developed, and of the place that system safety has attained. Authors don't agree on when or where it all started. But all the historical references on system safety do relate to the military or to aeronautics.

There are many citations in the literature of adverse accident experience in the military branches, which are said to have given impetus to the development of system safety concepts. This is an example taken from "Why 'System Safety'?" (7) by Charles O. Miller, a former director of the Bureau of Aviation Safety at the National Transportation Safety Board.

> Statistics show that far more aircraft, and indeed more pilots, were lost in stateside operations during [World War II] than ever were in combat. In 1943, for example, something like 5,000 aircraft were destroyed stateside, against 3,800 in the war proper.
>
> Then, shortly after World War II, in the period 1946 to 1948, people in the military were astonished by a new accident peak.
>
> Thus, the war experience, plus the immediate postwar experience, resulted in a call to the technical community for help.

But, in his *Handbook of System and Product Safety* (8), Willie Hammer says that:

> Oddly enough, it was more the concern with unmanned systems, the intercontinental ballistic missiles (ICBMs), that led to the development of the system safety concept.

C. W. Childs, in "Industrial Accident Prevention Through System Safety" (9), gives this brief history of the origins of system safety.

> In the early stages of missile development, it was necessary to assemble thousands of subsystems and millions of component parts in such a manner as to be virtually free or at least "forgiving" of mistakes or failures. The management and engineering disciplines at this point in time were not sophisticated or rigid enough to assure such a concept.
>
> Consequently, there were some very spectacular accidents and the engineering disciplines of reliability and quality control were tasked to provide a greater measure of mission success through elaborate quality assurance and reliability analyses and testing. Despite this, accidents continued at a higher rate than was considered acceptable and the new discipline of system safety was developed to bring the accident preventive experience into the systems engineering processes.
>
> Since that time, some methods and techniques have been developed which not only resulted in decreased accident rates for complex flight systems, but are now being used to prevent accidents in the industrial environment.

And great successes were achieved. But, I do wish that Childs' latter statement—that system safety concepts which were applied in developing complex flight systems were being used to prevent accidents in the industrial environment—could be substantiated more than I have been able to do.

There were many developments to advance system safety in the military branches and at the National Aeronautical and Space Administration (NASA) in the '60s, '70s, and '80s. Several standards were issued on system safety, the most prominent being *Military Standard—System Safety Program Requirements*—MIL-STD-882C (10). Its most recent modification was made in 1993.

Why System Safety Concepts Have Not Been Widely Adopted

At least one other author expected a more widespread adoption of system safety concepts, beyond the use by the military and aerospace personnel, and nuclear facility designers. He also had to recognize that it wasn't happening. In *The Loss Rate Concept in Safety Engineering* (11), R. L. Browning wrote this:

> As every loss event results from the interactions of elements in a system, it follows that all safety is "systems safety." . . . The safety community instinctively welcomed the systems concept when it appeared during the stagnating performance of the mid-1960s, as evidenced by the ensuing freshet of symposia and literature. For a time, it was thought that this seemingly novel approach could reestablish the continuing improvement that the public had become accustomed to; however, this anticipation has not been fulfilled.

> Now, almost two decades later, although systems techniques continue to find application and development in exotic programs (missiles, aerospace, nuclear power) and in the academic community, they are seldom if ever met in the domain of traditional industrial and general safety.

Although there were countless seminars and a proliferation of papers on system safety, the generalist in the practice of safety seldom adopted system safety concepts. In response to his own question—Why this rejection?—Browning said that traditional safety is predicated on absolutes, "safe" or "unsafe," while the concepts of measurable and acceptable risk are fundamental to system safety. Also, he expressed the view that system safety literature and seminars on system safety may have turned off generalist safety professionals because of the "exotica" they usually presented.

In *MORT Safety Assurance Systems* (1), Johnson cited two obstacles that deter the adoption of system safety concepts in industrial safety practice: the non-continuous nature of system safety work, being project-oriented; and, applying system safety methods to other than significant hazards can be "overkill."

Nevertheless, Browning went on to build *The Loss Rate Concept of Safety Engineering* (11) on system safety concepts. He also gave this encouragement:

> . . . we have found through practical experience that industrial and general safety can be engineered at a level considerably below that required by the exotics, using the mathematical capabilities possessed by average technically minded persons, together with readily available input data.

And it is appropriate to recognize that system safety concepts were foundational in the development of The Management Oversight and Risk Tree—MORT. References to system concepts are frequent in Johnson's text and in other literature on MORT.

There is a reality in the Browning and Johnson observations: system safety literature is loaded with governmental jargon, and can easily repel the uninitiated. It makes more of highly complex hazard analysis and risk assessment techniques requiring extensive knowledge of mathematics and probability theory than it does of concepts and purposes.

Some system safety literature does give the appearance of exotica. And using some of the methodologies for the analyses of hazards of lesser significance would be cost prohibitive.

Promoting the Use of System Safety Concepts

With the hope of generating a further interest by generalist safety professionals in system safety, I suggest that they concentrate on those basic system safety concepts through which gains can be made in an occupational setting, and avoid being repulsed by the more exotic hazard/risk assessment methodologies. Ted Ferry said it well in a preface to *System Safety 2000* (3) by Joe Stephenson:

> Professional credentials or experience in "system safety" is not required to appreciate the potential value of the systems approach and system safety techniques to general safety and health practice.

To paraphrase Browning, all hazards-related incidents result from interactions of elements in a system. Therefore, all safety is system safety. Therein lies an important idea.

Others have said that the system is what's important and that system safety purposes could be met, in many instances, with sound knowledge of safety practice, and intuition.

In *Safety Management* (12), John V. Grimaldi and Rollin H. Simonds wrote that:

> System safety analyses require the imaginative construction of every conceivable situation that could arise.
>
> A reference to system analysis may merely imply an orderly examination of an established system or subsystem.

Applying system safety as "an orderly examination of an established system or subsystem" to identify, analyze, avoid, eliminate, or control hazards can be successful in the less complex situations, without using elaborate analytical methods.

For safety generalists who take an interest in system safety concepts, I offer this short reading list.

Recommended Reading

- As a primer, *Military Standard—System Safety Program Requirements* (MIL-STD-882C) (10)
  System safety concepts and program requirements, and task descriptions for analytical techniques are included.
- Also as a primer, an article by Pat Clemens titled "A Compendium of Hazard Identification & Evaluation Techniques for System Safety Application" (13), in which comments are made on twenty-five analytical methods.
- *System Safety 2000* (3) by Joe Stephenson
  This book begins with a history of the fundamentals of system safety. Then the author moves into system safety program planning and management, and system safety analysis techniques. About half of the book is devoted to those techniques. A safety generalist would find it a good and not too difficult read.

- *Basic Guide to System Safety* (14) by Jeffrey W. Vincoli
  These two sentences are taken from the preface. "It should be noted from the beginning that it is not the intention of the *Basic Guide to System Safety* to provide any level of expertise beyond that of novice. Those practitioners who desire complete knowledge of the subject will not be satisfied with the information contained on these pages." Vincoli fulfilled his purpose. He has written a basic book on system safety that will serve the novice well.
- *System Safety Engineering and Management* (4) by Harold E. Roland and Brian Moriarty
  An overview of a system safety program is given, and descriptions of several analytical techniques. This is a good book: it covers system safety extensively.
- *Handbook of System and Product Safety* (8) by Willie Hammer
  This is one of the first texts on system safety and its methods. For quite some time, it has been a good reference.
- *The Loss Rate Concept In Safety Engineering* (11) by R. L. Browning
  This is a good little book, to which I have referred several times. Browning believes that one can apply system safety concepts in an industrial setting without necessarily becoming exotic. He builds on the Energy Cause Concept, and works through qualitative and quantitative analytical systems.
- *Managing Risk: Systematic Loss Prevention for Executives* (15) by Vernon L. Grose
  A system approach is taken by Grose for hazard identification, the writing of scenarios concerning them, and judgments made by "juries" of qualified personnel that consider the scenarios for probability of occurrence, severity of outcome, and the cost of risk control. Rankings, which are non-numerically quantified, are given to risks according to a Hazard Totem Pole. Grose is leery of the "numerologists." Reading this book is a good learning experience.

In this chapter, it was my purpose to recognize the many successes that have been achieved through the application of system safety concepts; establish that fundamental system safety concepts can be applied by generalists in safety practice; outline the "System Safety Idea"; encourage generalists who have not adopted system safety concepts in their practices to commence the inquiry and education to do so; and provide an introduction to Chapter 12, "Safety Professionals and the Design Processes."

I sincerely believe that we generalists in safety practice can learn from system safety successes and be more effective in our work through adopting system safety concepts. Their application in the occupational setting would result in significant reductions in injury and illness incidents.

References

1. Johnson, William G. *MORT Safety Assurance Systems.* New York: Marcel Dekker, 1980.
2. *Scope and Functions of the Professional Safety Position.* Des Plaines, IL: American Society of Safety Engineers, 1993.
3. Stephenson, Joe. *System Safety 2000.* New York: Van Nostrand Reinhold, 1991.
4. Roland, Harold E., and Brian Moriarty. *System Safety Engineering and Management.* New York: John Wiley & Sons, 1983.
5. Pearson, Richard G., and Mahmoud A. Ayoub. "Ergonomics Aids Industrial Accident and Injury Control." *IE* (June 1975).
6. "Search I—Fourth Installment: Locating Hazards Before They Become Accidents." *Journal of the American Society of Safety Engineers* (May 1970).
7. Miller, Charles O. "Why 'System Safety'?" *Technology Review* (February 1971).
8. Hammer, Willie. *Handbook of System and Product Safety.* Englewood Cliffs, NJ: Prentice-Hall, 1972.
9. Childs, C. W. "Industrial Accident Prevention Through System Safety." *Hazard Prevention* (September/October 1974).
10. *Military Standard—System Safety Program Requirements,* MIL-STD-882C. Department of Defense, Washington, D.C., 1993.
11. Browning, R. L. *The Loss Rate Concept in Safety Engineering.* New York: Marcel Dekker, 1980.
12. Grimaldi, John V., and Rollin H. Simonds. *Safety Management.* Homewood, IL: Irwin, 1989.
13. Clemens, P. L. "A Compendium Of Hazard Identification & Evaluation Techniques for System Safety Applications." *Hazard Prevention* (March/April 1982).
14. Vincoli, Jeffrey W. *Basic Guide to System Safety.* New York: Van Nostrand Reinhold, 1993.
15. Grose, Vernon L. *Managing Risk: Systematic Loss Prevention for Executives.* Englewood Cliffs, NJ: Prentice-Hall, 1972.

Chapter 12

Safety Professionals and the Design Processes

Introduction

To be effective, safety professionals must influence the design processes. Every chapter in this book speaks of the implications of design and engineering within the practice of safety. We are compelled, as professionals, to take into account the recognition that engineering is the preferred course of action to avoid, eliminate, or control hazards.

For all but a few safety professionals, what I propose is an entirely new venture—that they undertake to influence design and purchasing decisions to avoid bringing hazards into the workplace. That requires taking an anticipatory approach to safety, rather than being reactive. This new venture spells opportunity. Design, as the term is used here, encompasses all processes applied in devising a system to achieve results.

For many years, I have believed that the greatest strides forward respecting safety, health, and the environment will be made in the design processes. Slowly, a greater awareness has emerged of the soundness of that premise. Recent developments lead to the conclusion that, over time, the level of safety achieved will relate directly to the caliber of the design of facilities, hardware, equipment, tooling, operations layout, the work environment, and the work methods.

Some History, and a View of the Future

When I first entered the safety profession, almost all of the work done was of an engineering nature and dealt primarily with the physical aspects of facilities and equipment. Quite often, the three E's—engineering, education, and enforcement—were cited in the literature as the foundation for the practice of safety. And engineering was quite prominent in what we did.

Then came the behaviorists and the management systems people, who have had a significant influence on the safety profession. Their premises are based on the belief that about 90 percent of all industrial accidents are caused primarily by the unsafe acts of employees. Some safety professionals, in their practices, give minimum attention to the influence of design and engineering decisions on incident causation.

Safety literature published in the last twenty years has contained infrequent references to the design and engineering aspects of hazards management. A few well-known writers would have you believe that behavior modification, training, and management systems (consisting largely of what is referred to in OSHA literature as administrative controls) are almost the entirety of the practice of safety. How absurd!

Until recently, design and engineering considerations had largely fallen out of our concerns, unless they were imposed on us by legislation.

Yet in one subset of hazards management, the principal emphasis has been on design and engineering, with great success. Fire protection engineers have achieved an earned recognition for their capabilities as design consultants and are often brought into discussions by architects and engineers seeking their counsel on design specifications. Obviously, then, it can be done. Others engaged in the practice of safety can learn from the successes of fire protection professionals.

I believe that W. Edwards Deming (1) got it right when he wrote that a large majority of the problems in any operation are systemic, deriving from the workplace and work methods created by management, and that responsibility for only the relatively small remainder lies with the worker. Thus safety professionals, to be effective, must have an impact on upstream decision-making that determines the initial design of the workplace and work methods, and on decisions that subsequently affect the redesign of systems.

Just how effective is a safety professional's work if it does not impact on system design or redesign and consists mostly of what OSHA refers to as administrative controls?

Robert Andres takes an appropriate and futuristic view of the role of the safety professional in designing for safety in these excerpts from ASSE's *Engineering Division News* (2).

> . . . if we are to reduce injuries further and properly utilize scarce resources—we must dig deeper—and think more logically. This is the mandate of our profession as we "engineer types" have always envisioned it. In designing machines and processes, manufacturers must gain greater insight into projected uses, and misuses, of equipment—including planned and unplanned maintenance. Hazard identification must recognize all the possible

> hazards of the system—from the time it comes off the truck, until it becomes scrap.
>
> In the next millennium, design and manufacturing engineers will have a stronger safety background—and surviving safety professionals will offer a whole lot more than blind acceptance of regulations and an ability to write reports. The vista for the true safety engineer is wide open!

For each of the three major elements in the practice of safety—pre-operational, operational, and post-incident—the impact of design decisions is significant.

Design Implications in the Pre-operational Stage

Opportunities are greatest in the pre-operational mode for hazard identification and analysis and for their avoidance, elimination, or control. In the design process, the goal is to avoid bringing uncontrolled hazards into the workplace. That presents much opportunity for upstream involvement by safety professionals who would influence those making design and purchasing decisions. Their activities would include providing design specifications, giving consultation to those who design on safety goals to be achieved, assisting in design reviews, and developing specifications for the purchase of new equipment.

Requirements to achieve an acceptable risk level in the design process can usually be met without great cost if the decision-making occurs sufficiently upstream. When that does not occur, and retrofitting to eliminate or control hazards is proposed, the cost may be so great as to be prohibitive. The result may be a situation in which the risks are judged to be excessive.

Joe Stephenson was correct when he wrote this in *System Safety 2000* (3):

> The safety of an operation is determined long before the people, procedures, and the plant and hardware come together at the work site to perform a given task.

It is a hard truth that most of the significant, work-related safety decisions are made in the design processes. That is why the emphasis given here is so strong in support of safety professionals taking an anticipatory and proactive approach to hazards to avoid their being brought into the workplace.

Design Implications in the Operational Mode

The goal in the operational mode is to eliminate or control hazards *before their potentials are realized and hazards-related incidents occur.* Achieving that goal should be undertaken within a continuous improvement process. For every safety initiative, the first consideration should be, Can hazards be eliminated or controlled through redesign of the workplace or work methods?

Examples of the type of counsel-giving activities by safety professionals for which redesign considerations should be paramount are: any discussion of hazards, task reviews, giving counsel to safety committees and inspection teams, and job hazard analyses.

A job hazard analysis is a job design review that is to assess the physical hazards and the task performance hazards, taking into consideration the capabilities and limitations of people, and their possible quirky behavior.

For work hazards that are not to be eliminated or controlled through a redesign initiative, obviously the appropriate administrative practices would be applied. In so doing, the procedures employed should keep the risks of employee injury or illness or environmental damage at an acceptable minimum.

Design Implications in the Post-incident Mode

The goal of investigations of hazards-related incidents is to identify and eliminate or control their causal factors. If that job is done properly, many of the causal factors identified will relate to workplace and work methods design problems. Benefits to be obtained through effective incident investigation are discussed in Chapter 9, "Designer Incident Investigation." Much is made in this book of not assuming—post-incident—that employee error, the so-called unsafe act, is the principal causal factor when the reality is that work methods design causal factors were primary.

Consider these cases. In each one, the initial conclusion was that the employee action was the principal causal factor.

Case 1

Because of a glitch in production scheduling, delivery of parts by a conveyor to a workstation ceased. The design of the conveyor was such that parts would regularly fall off and accumulate beneath it. An employee, wanting to keep up with production needs, went beneath the conveyor to retrieve the parts that had collected there. Her hair was caught in a drive belt, and part of her scalp was torn away.

There were highly emotional meetings during which line employees were cautioned that they were not to enter the space beneath the conveyor. At first, the causal factor for this incident was recorded as the unsafe act of the employee. Subsequently, questions were posed about the significance of the production scheduling glitch and the design of the conveyor. If parts had not fallen off the conveyor, there would have been no enticement for the worker to retrieve them. The design of the conveyor was modified.

Case 2

After several fatiguing hours of work, an employee chose not to follow the established procedure to lock out and tag out the electrical power during a maintenance operation. He was electrocuted. The incident investigation report recorded the causal factor as "employee failed to follow the established procedure." Later, it was determined that the distance to the power shutoff was 216 feet. In that work situation, where fatigue had become a factor, it was judged that the system design "encouraged" the employee's risky behavior.

Case 3

Packages weighing in excess of twenty-five pounds were handled by workers several hundred times a day. Operations involved twisting, turning, and bending. Numerous workers compensation claims had been filed for back injuries. Assumptions were made that "improper lifting" was the causal factor needing attention. Over time, focused training was given and back belts were used.

Not until automatic tilt tables were provided, along with other steps to redesign the work, were the ergonomic stresses—and the workers compensation cases—reduced.

Training and personal protective equipment were not proper solutions for this situation, and the temporary "fixes" cost far more than would have been expended if the issues were addressed in the original design processes.

A Macro View of Design Implications

Although the scenarios just given are post-incident, their purpose is to develop an awareness of the workplace or work methods design implications on safety, and of the benefits that could be achieved if hazards are properly addressed in the design processes.

If the design of the work is overly stressful or error provocative, or if the immediate work situation encourages riskier actions than the prescribed work methods, the causal factors are principally systemic. To identify the causal factor in such situations primarily as an "employee error" or as an "unsafe act" would be wrong, inappropriate, and ineffective.

Consider a bit further how designing for safety relates on a macro basis to the reality of injury and illness experience. To demonstrate the relevancy, excerpts have been taken from a U.S. Department of Labor study of the characteristics of injuries and illnesses resulting in absences from work and shown in Table 12.1 (4).

For every event or exposure type listed in this exhibit, proper design practices would be beneficial in achieving injury and illness reduction.

Table 12.1. United States Department of Labor, Bureau of Labor Statistics Characteristics of Injuries and Illnesses Resulting in Absences from Work, 1994

This study was based on 2.25 million cases that resulted in at least a day away from work beyond the day of injury or onset of illness. Total and broad event or exposure categories may include data for classifications in addition to those shown separately. (Because of rounding and classifications not shown, percentages may not add up to 100.)

Event or Exposure	Percent distribution, 1994 cases	
Bodily reaction and exertion	44	
Overexertion		27
Bodily reaction, e.g., slip, fall		11
Repetitive motion		4
Contact with objects and equipment	27	
Struck by object		13
Struck against object		7
Caught in or compressed by equipment or objects		4
Falls	18	
Falls on same level		12
Falls to lower level		5
Exposure to harmful substance or environment	5	
Transportation accidents	4	
Assaults and violent acts	1	
Assaults by person(s)		1
Total	99	

Source: United States Department of Labor, Bureau of Labor Statistics (USDL-96-163), May 8, 1996.

Benefits to Be Obtained in Addition to Safety

Many companies have applied the principles of safety through design and their successes have impacted favorably on productivity and cost efficiency, in addition to safety. Their successes became evident in the work that culminated in the creation of an Institute for Safety through Design by the National Safety Council. This is the Institute's mission statement:

> To reduce the risk of injury, illness, and environmental damage by integrating decisions affecting safety, health, and the environment in all stages of the design process.

For the preparation of the proposal recommending that an Institute for Safety through Design be created, a collection was made of actual cases describing initiatives taken by safety professionals to resolve injury problems by redesigning operations that also resulted in improvements in productivity and cost efficiency. An overwhelming number of examples were submitted of pre-operational, operational, and post-incident situations. Several of those cases are cited in Chapter 10, "Applied Ergonomics: Significance and Opportunity."

As the brochure of the Institute for Safety through Design indicates, if hazards are properly addressed in the design processes, the benefits to society would be:

- significant reductions of injuries, illnesses, and damage to the environment, and their attendant costs;
- productivity improvement;
- decreases in operating costs; and
- the avoidance of design shortcomings and expensive retrofitting

A good example of the benefits that can be achieved from applying a safety through design concept is described in these excerpts from an article titled "Sound Advice for a Quieter Workplace" that appeared in the magazine *Safety+Health* (5). For this article the subcaption is "Companies take a proactive stance against noise with engineering controls."

> Over months and years, the total costs of replacing equipment and the annual testing of every exposed employee can become significant, both in dollars and in time. Finally, when everyone is wearing hearing protection, *errors and delays in communication might cause costly losses in quality and productivity.* [Italics added.]
>
> The limited effectiveness of administrative controls and testing has led many companies to the conclusion that engineering controls offer the most reliable and lasting way to reduce workplace noise.
>
> The most effective time to apply engineering controls is before a noise source ever comes into the workplace. Establish noise standards and work closely with suppliers at the time machinery is specified and ordered. Ford's noise standard, SX-1, has been in effect for more than 20 years and is routinely included in the specifications for new and rebuilt machinery.
>
> [When a Ford plant was reconstructed] noise was eliminated by design. . . . Ford's efforts have paid off. Recent tests that equipped 300 employees in the plant with noise dosimeters showed none were being exposed to noise above 85 dBA. Of 700 employees in the new plant, fewer than 60 are required to wear hearing protection or undergo testing. . . . any company that buys equipment can specify maximum noise levels for new purchases.

Being Anticipatory and Proactive

During my discussions with Dr. Thomas A. Selders in developing a definition of the practice of safety, he offered this critique: the principal shortcoming in what safety professionals do is that they seldom are in a position to anticipate hazards and give counsel on their avoidance. Dr. Selders' point was that our activity did not start soon enough in the decision stream, that it was not proactive.

Willie Hammer addresses the need to anticipate, avoid, or control hazards in the design process in his *Handbook of System and Product Safety* (6). This is what he writes.

> The system safety (and product safety) concept is predicated on this principle: The most effective means to avoid accidents during system operation is by eliminating or reducing hazards and dangers during design and development.

Hammer's statement applies to every aspect of safety, whatever it is called. His premise spells opportunity for safety professionals to provide counsel in the design process on a proactive basis, to anticipate hazards and to give advice on their avoidance, elimination, or control.

On "Human Error" and Designing for Safety

For many incidents resulting in back injuries, strains and sprains, cumulative trauma disorders, and others, employees were adhering to the prescribed work procedure when the incident occurred, but typically, an incident investigation report would show that an unsafe act, a human error, was the principal incident cause.

If the employee was following the prescribed procedure, and did not commit an error, surely the focus in determining causal factors should be on the design of the prescribed work methods. It would be inappropriate in such a situation to indicate that the principal cause of the incident was an employee unsafe act, when the employee was doing precisely what was expected.

One of my purposes here is to suggest that, often, a causal factor identified as an "employee error" or an "unsafe act" may actually be "programmed" into the prescribed work methods. That occurs when the design of the work is overly stressful, or error provocative, or encourages riskier actions than desired. If the work is so designed, it is reasonable to assume that a performance deviation is principally a systemic problem rather than a task performance problem.

Alan D. Swain, in a paper titled "Work Situation Approach to Improving Job Safety" (7), suggested that management "forego the temptation to place the burden of accident prevention on the individual worker." This is an excerpt from that paper:

> . . . a means of increasing occupational safety is one which recognizes that most human initiated accidents are due to the features in a work situation which define what the worker must do and how he must do it . . . . the situation approach emphasizes structuring or restructuring the work situation to

prevent accidents from occurring. Use of this approach requires that management recognize its responsibility (1) to provide the worker with a *safety-prone* work situation and (2) to forego the temptation to place the burden of accident prevention on the individual worker.

Alphonse Chapanis is exceptionally well-known in ergonomics and human factors engineering circles. He is often quoted concerning the gains to be made by addressing the capabilities and limitations of workers in the design processes, thereby avoiding the design of work that is error-provocative.

These are excerpts from his chapter "The Error-Provocative Situation" in *The Measurement of Safety Performance* (8).

- Many situations are error provocative. . . . the evidence is clear that people make more errors with some devices than they do with others.
- Given a population of human beings with known characteristics, it is possible to design tools, appliances, and equipment that best match their capacities, limitations and weaknesses.
- The improvement in system performance that can be realized from the redesign of equipment is usually greater than the gains that can be realized from the selection and training of personnel.
- Design characteristics that increase the probability of error include a job, situation, or system which:
  a. Violates operator expectations;
  b. Requires performance beyond what an operator can deliver;
  c. Induces fatigue;
  d. Provides inadequate facilities or information for the operator;
  e. Is unnecessarily difficult or unpleasant; or
  f. Is unnecessarily dangerous.
- . . . a good systems engineer can usually build a nearly infallible system out of components that individually may be no more reliable than a human being. The human factors engineer believes that with sufficient ingenuity nearly infallible systems can be built even *if* one of the components is a human being.

A central point in Dr. Chapanis' work is that: "The improvement in system performance that can be realized from the redesign of equipment is usually greater than the gains that can be realized from the selection and training of personnel."

Trevor Kletz, in his book *An Engineer's View of Human Error* (9), states that:

> Almost any accident can be said to be due to human error and the use of the phrase discourages constructive thinking about the action needed to prevent it happening again; it is too easy to tell someone to be more careful.

Kletz also suggests that we should do away with the term "human error." His focus, in his introduction, has a significant bearing on the purpose of this chapter:

> The theme of this book is that it is difficult for engineers to change human nature and, therefore, instead of trying to persuade people not to make mistakes, we should accept people as we find them and try to remove opportunities for error by changing the work situation, that is, the plant or equipment design or the method of working. Alternatively, we can mitigate the consequences of error or provide opportunities for recovery.
>
> A second objective of the book is to remind engineers of some of the quirks of human nature so that they can better allow for them in design.

In his book, Kletz reviewed several accidents that "at first sight were due to human error" and discussed how, in reality, they could have been prevented through improved design, construction, and maintenance, and through improved design of work methods, and better management.

I now conclude that an engineer who is to design the workplace and work methods must, I repeat, must, be learned in ergonomics. Why? Without a thorough understanding of the capabilities and limitations of people, and their penchant for unpredictable behavior, the design engineer could not *effectively anticipate stressful or error-provocative behavior—or quirky behavior—and design systems that compensate for, or are tolerant of, possible human error.*

Having so stated, I want to make sure that I do not imply that designing for safety is limited to ergonomics. Designing for safety applies to all injury and illness types.

Designing for Safety during Maintenance

My experience over many years has been that, in relation to their share of the employment population, maintenance workers had higher lost work day case rates (i.e., severe injuries). To determine whether others had similar experiences, I polled twenty-one safety directors for their comments: eighteen said that their companies had similar results; three said that theirs did not.

Nearly all of those eighteen safety directors had their own tragic stories to tell about maintenance people having incidents resulting in severe injuries. And in each case, questions could be asked about the absence of safety considerations in the design of operations that resulted in maintenance work being so risky.

Not one of us could say that we had historically done a good job of exploring the implications of workplace or work methods design during investigations of incidents involving maintenance personnel. Nor could one of those eighteen safety directors recall a helpful study on the subject. Yet all of us recognized the significance of the higher lost workday case rates that maintenance personnel have had.

A limited search for literature that addresses designing for safety during maintenance came up short. There is a relevant paragraph or two in some books on maintenance or maintainability. This subject should be given a high priority so that the knowledge required about designing for safety during maintenance is developed, published, and promoted.

CONCLUSION

Frequent opportunities can be created by safety practitioners to influence the design processes, to their great advantage. To do that effectively requires:

- application of an incident causation model that properly balances causal factors deriving from less than adequate policies, standards, or procedures that impact on design management, operations management, and task performance;
- an understanding of hazards;
- an understanding of risk;
- knowledge of hazard analysis and risk assessment methods;
- awareness of designing for safety concepts;
- ability to demonstrate to senior management, design personnel, and purchasing personnel the benefits to be obtained from designing at minimum risk.

The next chapter is to assist safety professionals who become involved in the design processes by providing guidelines for designing for safety.

REFERENCES

1. Deming, W. Edwards. *Out of the Crisis.* Cambridge, MA: Center for Advanced Engineering Study, Massachusetts Institute of Technology, 1986.
2. "Engineering News Letter." American Society of Safety Engineers, Des Plaines, Il; Summer 1996.
3. Stephenson, Joe. *System Safety 2000.* New York: Van Nostrand Reinhold, 1991.
4. "Characteristics of Injuries and Illnesses Resulting in Absences From Work, 1994." USDL-96-163. U.S. Department of Labor, Bureau of Labor Statistics, Washington, D.C., May 8, 1996.

5. "Sound Advice for a Quieter Workplace," *Safety+Health.* Itasca, IL: National Safety Council, February, 1996.
6. Hammer, Willie. *Handbook of System and Product Safety.* Englewood Cliffs, NJ: Prentice-Hall, 1972.
7. Swain, Alan D. "Work Situation Approach to Improving Job Safety." Sandia Laboratories, Albuquerque, NM.
8. Chapanis, Alphonse. "The Error-Provocative Situation." In *The Measurement of Safety Performance,* ed. William E. Tarrants. New York: Garland Publishing, 1980.
9. Kletz, Trevor. *An Engineer's View of Human Error.* Rugby, Warwickshire, UK: Institution of Chemical Engineers, 1991.

Chapter 13

Guidelines: Designing for Safety

Introduction

Assume that a safety professional wants to influence an organization's culture as it impacts on the design of the workplace and work methods. A first step might be to search for readily available guidelines on the concept of designing for safety: that search may not be productive.

I have not located a resource that sets forth in concept form the principles to be applied in designing for safety. For design purposes, standards, regulations, specifications, design handbooks, and checklists that establish the minimums for specific design subjects are plentiful, and they are very important. But a concept of designing for safety must go far beyond the application of standards and guidelines.

Consider this example of a concept that goes far beyond standards and guidelines: a highly significant goal to be accomplished in designing for safety is that a "user-friendly" occupational setting is to be achieved. That is the subject of *Plant Design for Safety: A User-Friendly Approach* (1) by Trevor Kletz. Consider his theme:

> In all industries errors by operators and maintenance workers and equipment failures are recognized as major causes of accidents, and much thought has been given to ways of reducing them or minimizing their consequences. Nevertheless, it is difficult for operators and maintenance workers to keep up an error-free performance all day, every day. We may keep up a tip-top performance for an hour or so while playing a game or a piece of music, but we cannot keep it up continuously.

> Designers have a second chance, opportunities to go over their designs again, but not operators and maintenance workers. Plants should therefore be designed, whenever possible, so that they are "user friendly," to borrow a computer term, so that they can tolerate departures from ideal performance by operators and maintenance workers without serious effects on safety, output, or efficiency.
>
> It is the theme of this book that, instead of designing plants, identifying hazards, and adding on equipment to control the hazards or expecting operators to control them, we should make every effort to choose basic designs and design details that are user friendly.

What needs to be agreed upon in an organization is a well-understood concept, a way of thinking, that is translated into a process that effectively addresses hazards and risks in the design processes, a way of thinking that is universally applied by all involved with designing.

Principal Resources

These guidelines are to provide safety professionals and design engineers with concepts to be applied in designing for safety. In their development, considerable extension was required of the bits and pieces taken from several sources, two of which are most significant.

One is *Military Standard System Safety Program Requirements,* often referred to as MIL-STD-882C (2). And the second is William J. Haddon's energy release theory as set forth in two of his papers, "The Prevention of Accidents" (3) and "On the Escape of Tigers: An Ecological Note" (4).

Military Standard 882C is the document containing the system safety program requirements for contractors to the Department of Defense. It is not written in a language easily transferable to the needs of the occupational safety professional or to design engineers responsible for the occupational setting. Nevertheless, I consider it to be a valuable resource. As I extract from it, I hope that safety generalists will agree that its premises are applicable to their fields of endeavor.

Haddon's energy release theory proposes that quantities of energy, means of energy transfer, and rates of transfer are related to the types of incidents that occur, the probability of their occurrence, and the severity of their outcomes. Haddon also addressed the significance of avoiding unwanted exposures to harmful environments. Designing to avoid unwanted energy flows and unwanted exposures to harmful environments should reduce both the probability of harmful or damaging incidents occurring, and the severity of their outcomes.

Using Haddon's theory as a base, considering comments of others on his work, and making extensions representing my thinking, "A Generic Thought Process for Hazard Avoidance, Elimination, or Control" was developed.

Ultimate Purpose

The ultimate purpose of applying concepts of safety through design to systems, the workplace, and work methods is to achieve safety—that state for which the risks are judged to be acceptable.

General Principles and Definitions

For the practice of safety, the term design processes applies to:

- facilities, hardware, equipment, tooling, selection of materials, operations layout and configuration, energy control, environmental concerns
- work methods and procedures, personnel selection standards, training content, work scheduling, management of change procedures, maintenance requirements, and personal protective equipment needs

There must be an understanding of risk by those who are involved in the design process. Risk is defined as a measure of the probability of a hazards-related incident occurring and the severity of harm or damage that could result.

As a matter of principle, for an operation to proceed, its risks must be acceptable. Risks are acceptable if they are judged to be tolerable.

Minimum risk is to be sought with respect to new technology, materials, and designs, and in designing new production methods.

Minimum risk is achieved when all risks deriving from hazards are at a realistic minimum. Minimum risk does not mean zero risk, which is unattainable.

In determining minimum risk, decision factors will be design objectives, the practicality of risk reduction measures and their costs, and their probable acceptance by users.

If a system—the facilities, equipment, and work methods—is not designed to minimum risk, superior results with respect to safety cannot be attained, even if management and personnel factors approach the ideal.

In the design and redesign processes, the two distinct aspects of risk must be considered:

- avoiding, eliminating, or reducing the *probability* of a hazards-related incident occurring;
- minimizing the *severity* of harm or damage, if an incident occurs.

All risks to which this concept applies derive from hazards. There are no exceptions.

Thus, hazards must be the focus of design efforts to achieve a state for which the risks are judged to be acceptable.

Hazards are most effectively and economically avoided, eliminated, or controlled in the design or redesign processes.

Both the technology and human activity aspects of hazards must be addressed. A hazard is defined as the potential for harm: Hazards include the characteristics of things and the actions or inactions or people.

If a hazard is not avoided, eliminated, or controlled, its potential may be realized, and a hazards-related incident may occur that has the potential to, but may or may not, result in harm or damage, depending on exposures.

With respect to workstations, tools, equipment, and operating methods, design and engineering applications are the preferred measures for preventing hazards-related incidents since they are more effective.

Hazard probability is defined as the aggregate of the likelihood that the potentials of hazards will be realized and that a hazards-related incident will occur. Hazard probability is to be described in probable occurrences per unit of time, events, population, items, or activity.

Hazard severity is defined as the aggregate of the worst credible outcomes of a hazards-related incident, considering the exposure.

Exposure includes the people, property, and the environment that could be harmed or damaged if a hazards-related incident occurs.

A risk assessment is an analysis that addresses both the probability of a hazards-related incident occurring and the expected severity of its adverse effects.

Objectives

Safety, consistent with goals, is to be designed into all systems, processes, the workplace, and the work methods in a proactive, cost-effective manner.

Risk assessment is to be an integral part of the design process.

A fundamental design purpose is to have processes and products that are error-proof, or error-tolerant.

Hazards must be identified and evaluated, and then avoided, eliminated, or controlled so that the associated risks are at an acceptable level *throughout the entire life cycle of processes, equipment, and products.*

Consideration is to be given early in the design process to the risks attendant in the eventual disposal of processes and products.

Requirements for minimum risk are to be established and applied in the acquisition or acceptance of new materials, technology, or designs, and prior to the adoption of new production, test, or operating techniques.

Actions taken to identify and eliminate hazards and to reduce their attendant risks to an acceptable level are to be documented.

Retrofit actions required to improve safety are to be minimized through the timely inclusion of safety features during research, technology development, and in purchasing and acquisition.

Simplicity of design is to be attained—the simplest design consistent with functional requirements and expected service conditions. Systems and equipment are to be capable of operation, maintenance, and repair in their operational environment by personnel with a minimum of training.

When design or work methods changes are made, a management of change system is to be in place that includes identification and analysis of hazards so that an acceptable risk level is maintained.

Significant safety data representing lessons learned are to be documented and disseminated to interested personnel.

Order of Design Precedence

To achieve the greatest effectiveness in hazard avoidance, elimination, or control, the following order of precedence is to be applicable in all design and redesign processes:

Top Priority: Design for Minimum Risk

From the very beginning, the top priority is that hazards are to be eliminated in the design process. If an identified hazard cannot be eliminated, the associated risk is to be reduced to an acceptable level through design decisions.

Second Priority: Incorporate Safety Devices

As a next course of action, if hazards cannot be eliminated or their attendant risks adequately reduced through design selection, the risks are to be reduced to an acceptable level through the use of fixed, automatic, or other protective safety design features or devices. Provisions are to be made for periodic functional checks of safety devices.

Third Priority: Provide Warning Devices

When identified hazards cannot be eliminated or their attendant risks reduced to an acceptable level through initial design decisions or through the incorporated safety devices, systems are to be provided that detect the hazardous conditions and include warning signals to alert personnel of the hazards.

Warning signals and their application are to be designed to minimize the probability of incorrect personnel reactions and shall be standardized within like types of systems.

Fourth Priority: Develop and Institute Operating Procedures and Training

Where it is impractical to eliminate hazards or reduce their associated risks to an acceptable level through design selection, incorporating safety devices, or warning devices, relevant operating procedures and training shall be used.

However, operating procedures and training, or other warning, caution, or written advisory forms are not to be used as the only risk reduction method for critical hazards. Acceptable procedures may include the use of personal protective equipment.

It should be understood that tasks and activities judged to be essential to safe operation may require special training and certification of personnel proficiency.

For many design situations a combination of these principles will apply. But a lower level of priority is not to be chosen until practical applications of the preceding level or levels are exhausted. First and second priorities are more effective because they reduce the risk by design measures that eliminate or adequately control hazards; third and fourth priorities rely on human intervention.

General Design Requirements: A Thought Process for Hazard Avoidance, Elimination, or Control

This outline pertains to all three elements of the practice of safety—the pre-operational mode, the operational mode, and the post-incident mode. Some aspects pertain to either or both unwanted energy flows and unwanted exposures to harmful environments. In offering this outline, it is strongly emphasized that:

- the Order of Design Precedence previously given is to prevail;
- ergonomics design principles are to apply, so that the work methods prescribed are not error-provocative or overly stressful; and that
- the two distinct aspects of risk are to be considered in the design and redesign processes:
  - avoiding, eliminating, or reducing the *probability* of a hazards-related incident occurring, and
  - minimizing the *severity* of harm or damage, if an incident occurs.

1. Avoid introduction of the hazard: prevent buildup of the form of energy or hazardous materials:
   - avoid producing or manufacturing the energy or the hazardous material
   - use material handling equipment rather than manual means
   - don't elevate persons or objects
2. Limit the amount of energy or hazardous material:
   - seek ways to reduce actual or potential energy input
   - use the minimum energy or material for the task (voltage, pressure, chemicals, fuel storage, heights)

- consider smaller weights in material handling
- store hazardous materials in smaller containers
- remove unneeded objects from overhead surfaces

3. Substitute, using the less hazardous:
    - substitute a safer substance for a more hazardous one; when hazardous materials must be used, select those with the least risk throughout the life cycle of the system
    - replace hazardous operations with less hazardous operations
    - use designs needing less maintenance
    - use designs that are easier to maintain, considering human factors
4. Prevent unwanted energy or hazardous material buildup:
    - provide appropriate signals and controls
    - use regulators, governors, and limit controls
    - provide the required redundancy
    - control accumulation of dusts, vapors, mists, et cetera
    - minimize storage to prevent excessive energy or hazardous material buildup
    - reduce operating speed (processes, equipment, vehicles)
5. Prevent unwanted energy or hazardous material release:
    - design containment vessels, structures, elevators, material handling equipment to appropriate safety factors
    - consider the unexpected in the design process, to include avoiding the wrong input
    - protect stored energy and hazardous material from possible shock
    - provide fail-safe interlocks on equipment, doors, valves
    - install railings on elevations
    - provide non-slip working surfaces
    - control traffic to avoid collisions
6. Slow down the release of energy or hazardous material:
    - provide safety and bleed-off valves
    - reduce the burning rate (using an inhibitor)
    - reduce road grade
    - provide error-forgiving road margins
7. Separate in space or time, or both, the release of energy or hazardous materials from that which is exposed to harm:
    - isolate hazardous substances, components, and operations from other activities, areas, and incompatible materials, and from personnel
    - locate equipment so that access during operations, maintenance, repair, or adjustment minimizes personnel exposure (e.g., hazardous chemicals, high voltage, electromagnetic radiation, cutting edges)
    - arrange remote controls for hazardous operations
    - eliminate two-way traffic
    - separate vehicle from pedestrian traffic
    - provide warning systems and time delays

8. Interpose barriers to protect the people, property, or the environment exposed to an unwanted energy or hazardous material release:
   - insulation on electrical wiring
   - guards on machines, enclosures, fences
   - shock absorbers
   - personal protective equipment
   - directed venting
   - walls and shields
   - noise controls
   - safety nets
9. Modify the shock concentrating surfaces:
   - padding low overheads
   - rounded corners
   - ergonomically designed tools
   - "soft" areas under playground equipment

Hazard Analysis and Risk Assessment

A hazard analysis and risk assessment method must be used in determining risk and the hazards management actions to be taken. A good hazard analysis/risk assessment model will enable decision-makers to understand and categorize the risks and to determine the methods and costs to reduce risks to an acceptable level. Chapter 15, "Hazard Analysis and Risk Assessment," addresses this subject.

Appropriation Requests, Project Reviews, Contract Specifications

An organization's appropriation proposal system for new projects or major alterations should include a design review procedure that includes hazard identification and analysis and risk assessment. One purpose of addressing hazards in the design process is to avoid retrofitting, which may be excessively costly or impossible when construction projects are in progress or new equipment does not meet the requirements of good hazards management.

When contract specifications clearly set forth safety requirements, the probability of that sort of problem occurring is reduced. Some safety professionals have convinced their managements that they should be participants in specification writing.

Having a well-crafted checklist for project reviews and for the development of contract specifications would be beneficial to engineers and safety professionals. A sample checklist follows. It is to serve as a guide only. For every organization, revisions should be made to suit that entity's particular needs.

Project Review—Contract Specification Check List

Walking/Working Surfaces

1. Will pedestrian aisles and forklift aisles be ample?
2. In the aisles, are people and vehicles adequately separated?
3. Will adequate exits and egress paths be provided?
4. Have logistics been studied to provide safe and efficient flow of people and materials?
5. Will the construction texture of walking surfaces be non-slip?
6. Will the floors be designed to stay dry?
7. Will water and process flows be designed to keep off the walkway?
8. Will the floors be sloped and drained?
9. Will utilities and other obstructions be routed off the walking surfaces?
10. Will the design allow future utility expansion without having to cross floors?

Mechanical Safety

11. Will machines be properly guarded?
12. Will machines be properly interlocked?
13. Will machines be properly equipped for connecting equipment or conveyors?
14. Will hot surfaces be insulated?

Electrical Safety

15. Will the electrical system meet OSHA/NEC standards?
16. Is the design flexible enough to safely allow future service expansion?
17. Will the design allow safe use of temporary systems?
18. Will grounding and proper fuses or circuit breakers be provided?
19. Will ground fault interruption be provided for wet areas?
20. Will hard wiring be provided where needed?
21. Will emergency power be provided for critical systems?

Fire Protection

22. Will National Fire Codes (NFC) and insurance requirements for fire walls, doors, exits be met?
23. Will the building have adequate external fire zones?
24. Will emergency vehicle access be adequate?
25. Will flame arresters be installed where needed on equipment vents?
26. Will fire hydrants be adequate?
27. Will small hose stand pipes be adequate?
28. Will sufficient hose racks be provided?
29. Will fire extinguishers be of appropriate types, and adequate?

30. Will risers and post valves be accessible?
31. Will the location of flammables be appropriate?
32. For flammables, will storage rooms and cabinets meet NFC and insurance requirements?
33. For flammable liquid dispensing, will grounding, bonding, and ventilation be adequate?
34. Will special fire suppression systems be provided?
35. Has containment of fire suppression water been addressed?
36. Will fire sensors, pull stations, and alarms be adequate?
37. Has the project been reviewed by insurance company personnel?

Emergency Safety Systems

38. Will reliable emergency power be provided for critical and life-support systems?
39. Will remote or self-actuating valves be installed where necessary?
40. Will emergency lighting and exit lighting be adequate?
41. Will emergency safety showers and eye wash stations be adequate, and properly placed?
42. Will adequate first aid stations, spill carts, and emergency stations be provided?

Chemical, Biological, Radiological

43. Have all materials in this category been identified and inventoried?
44. Have the physical properties of the individual chemicals been identified?
45. Are the reactive properties known for chemicals that will be combined or mixed?
46. Have Material Safety Data Sheets been obtained for all materials?
47. Have adequate provisions been made for chemical release, fire, explosion, reaction?
48. Have measures been taken to minimize the quantities of hazardous chemicals stored?
49. Is the storage of hazardous chemicals below ground to be avoided?
50. Is storage tank location such as to minimize facility damage in a catastrophic event?
51. Is adequate storage tank diking provided?
52. Will emergency ventilation be provided for accidental releases?
53. For extraordinary releases, will special ventilation, relief, and deluge systems be provided?
54. Will the normal use of chemicals allow operating without personal protective equipment?
55. Will the design of bulk loading/unloading facilities contain anticipated leaks and spills?

Pressure Vessels

56. Will pressure vessels be designed to ASME and insurance company requirements?
57. Will vessels containing flammables or combustibles meet OSHA (1910:106) and NFC standards?
58. Will pressure relief valves be correctly sized and set, and suitable for intended use?
59. Will discharge of relief devices be directed safely?
60. Will needed vacuum breakers be installed?
61. Are products compatible with the composition of the vessels?

Ventilation

62. Will local ventilation effectively capture contaminants at the point of discharge?
63. Will room static pressures be progressively more negative as the operation becomes "dirtier"?
64. Will ventilation system provide a margin of safety if the system fails?
65. Will emergency power be provided on critical units?
66. Will the ventilation equipment be remote and/or "quiet"?
67. Will spray booths and degreasers meet OSHA standards?
68. Will laboratory or contaminated air be totally exhausted?
69. If contaminated air is cleaned and reused, is its safety assured?
70. Will the make-up air to hoods be adequate?
71. Have flow patterns been established to prevent exposure to personnel?
72. Will a glove box or process containment be provided for carcinogens or mutagens?
73. Is an adequate facility provided for radiation or biological hazards?
74. Does the ventilation meet American Society of Heating, Refrigerating and Air-Conditioning Engineers (ASHRAE) standards?

Ergonomics—Workstation and Work Methods Design

75. Do material handling designs consider employee capabilities and limitations?
76. Do material handling designs promote the use of hoists, scissor jacks, or drum carts?
77. Has constant lifting been eliminated or minimized in work methods design?
78. Have routine operations been designed to avoid crawling, stooping, or overreaching?
79. Has the need for stairs or ladders been reduced or eliminated?
80. Does the design safely accommodate routine servicing and cleaning?
81. Will there be adequate clearance around equipment for servicing?

82. Will controls be efficiently located in a logical and sequential order?
83. Will indicators be easy to read, either by themselves or in combination with others?
84. Has adequate attention been given to lighting, heat, cold, noise, and vibration?

Environmental

85. Have waste products been identified and a means of disposal established?
86. Will provisions be made for responding to chemical spills (containment, cleanup, disposal)?
87. Is there an existing spill control plan for chemicals?
88. Have all waste streams been identified?
89. Are adequate pretreatment facilities provided for process waste streams?
90. Will an adequate storage area be available for wastes held prior to treatment or disposal?
91. Will waste storage areas have adequate isolation or containment for spills?
92. Will hazardous wastes be disposed of at approved treatment, storage, and disposal facilities?
93. Is special equipment or specially trained personnel provided for treatment operations?
94. Has the acquisition of permits been addressed for the treatment or disposal of waste streams?
95. Have state or local requirements for permitting been evaluated and factored into the project?
96. Can the facility meet regulations for reporting spills or the storage of chemicals?
97. Have adequate provisions been made for cleaning the process equipment?
98. Have provisions been made for a catastrophic release of chemicals?
99. Have provisions been made for any necessary demolition and the resulting waste?
100. Have requirements for remediation at the site prior to construction been addressed?
101. Will all feasible measures for waste minimization be implemented?
102. Have the processes that generate air pollution been evaluated for minimization potential?
103. Will adequate air pollution controls be installed (scrubbers, fume hoods, dust collectors)?
104. Has handling and cleaning of air pollution control systems been addressed?

Wastewater

105. Have the processes that generate wastewater been evaluated for minimization potential?
106. Will indoor spills be protected from reaching drains?
107. Will outdoor spills be protected from reaching storm water drains and sewer manholes?
108. Are adequate water disposal systems available?
109. Will pretreatment methods be necessary and provided?
110. Will the discharges of domestic and industrial wastewater be in accord with regulations?

A Model Policy/Procedure Statement on Safety through Design

In a few organizations, policy/procedure statements have been issued specifying that hazards are to be addressed early on by design work groups, with safety professionals being participants in design discussions.

To serve as a reference, an example of a policy/procedure statement on safety in the design process follows. Its intent is to announce that hazards are to be addressed during early design concept stages, and as an integral part of a concurrent engineering program.

Safety in the Design Process
An Example of a Policy/Procedure Statement

It is our continuing policy to provide employees with a safe work environment, and to assure a proper treatment of environmental hazards deriving from our operations.

To meet this objective, it is necessary for personnel having design responsibilities to consider hazards during the early concept stages when developing new products, manufacturing processes, technology, and facilities that may impact on occupational safety and health and on the environment.

It is most cost-effective to design for safety, health, and environmental considerations upstream where the ability to influence is greatest. In addition to reducing risk, the concept of "Safety in the Design Process" has also been demonstrated to:

- Increase worker productivity
- Improve people and processing flexibility
- Facilitate uptime
- Reduce costs
- Reduce hazards in service and maintenance activities
- Achieve effective environmental controls, upstream

Conversely, the cost of secondary engineering to retrofit for hazards impacting on safety, health, and the environment after the initial design and deployment of the manufacturing process is excessive, and often includes burdensome constraints on our manufacturing and production systems.

During the early conceptual stages for product and process development, anticipating service and maintenance tasks and identifying employee exposures are critical first steps in developing the safeguarding and engineering controls necessary for protecting the employee. That same concept applies for environmental controls. It includes designing to avoid or control hazards and designing in the necessary safeguarding protection for operators and supporting maintenance personnel—considering both planned and unplanned service of the equipment and facility.

Engineering design should strive for elimination of hazards. Only when elimination, substitution, or engineering controls are not feasible should reliance on physical barriers, warning systems, training, and personal protective equipment be considered.

The concept of Safety in the Design Process requires a coordinated effort between the engineering and the safety, health, and environmental communities. Our company bulletin XYZ establishes when safety, health, and environmental studies are necessary in the consideration of new products, technology, and manufacturing processes. Please review current and future programs to assure that safety, health, and environmental issues are considered in the early stages of concept and design.

Safety, health, and environmental personnel are to assist as technical resources in achieving our Safety in the Design Process goals.

A Model Procedure Guide for Safety through Design

How would an organization put into practice a policy requiring that hazards be addressed in the design process? As was the case with policy statements, very few procedure statements exist that could serve as references. An adaptation follows of such a procedure guide. It is to serve as a reference only; it will not be suitable without modification for any organization.

Process Design and Equipment Review
An Example of a Procedure Guide

Purpose

To provide operations, engineering, and design personnel with guidelines and methods to foresee, evaluate, and control hazards related to occupational safety and health and the environment when considering new or redesigned equipment and processes.

Scope and Definitions

This guideline is applicable to all processes, systems, manufacturing equipment, and test fixtures regardless of size or materials used. These conditions will be necessary for an exemption from design review:

- no hazardous materials are used (as defined by 29CFR 1910.1200);
- operating voltage of equipment is less than 15 volts and the equipment will be used in non-hazardous atmospheres and dry locations;
- no hazards are present that could cause injury to personnel (e.g., overexertion, repetitive motion, error-prone situations, falls, crushing, lacerations, dismemberment, projectiles, visual injury, et cetera);
- pressures in vessels or equipment are < 2 psi;
- operating temperatures do not exceed 100°F/38°C;
- no hazardous wastes as defined by 40CFR 26 and 262 and/or 331 CMR 30 are generated;
- no radioactive materials or sealed source devices are used.

If other exemptions are desired, they are to be cleared by the safety, health, and environmental professional.

Phase I: Pre-Capital Review

This review is to be completed prior to submission of a project request or a request for equipment purchase, in accord with the capital levels outlined in Bulletin 246. Pre-capital reviews are crucial for planning facilities needs such as appropriateness of location, power supply, plumbing, exhaust ventilation, et cetera. Process and project feasibility are determined through this review. A complete "What If" hazard analysis, in accord with Bulletin 135, is to accompany the request. Non-capital projects should also be reviewed utilizing these procedures, but a formal "What If" hazard analysis is not required.

Phase II: Installation Review

This review requires a considerably more detailed hazards and failure analysis relative to equipment design, production systems, and operating procedures. Detailed information is documented, including equipment operating procedures, a work methods review giving emphasis to ergonomics, control systems, warning and alarm systems, et cetera. A "What If" system of hazard analysis may be used and documented. Other methods of hazards analysis will be applied if the hazards identified cannot be properly evaluated through the "What If" system. The project manager shall be responsible for establishing a hazard review committee and for managing its functions.

Hazard Review Committee

This committee will conduct all phases of design review for equipment and processes. In addition to the project manager, members will include the safety, health, and environmental professional, the facilities engineer, the design engineer, the manufacturing engineer, and others (financial, purchasing) as needed. For particular needs, outside consultants for equipment design or hazard analysis may be recommended by the safety, health, and environmental professional.

"What If" Hazard Analysis

This hazard assessment method utilizes a series of questions focused on equipment, processes, materials, and operator capabilities and limitations, including possible operator failures, to determine that the system is designed to a level of acceptable risk. Users of the "What If" method would identify possible unwanted energy releases or exposures to hazardous environments. Bulletin 135 contains procedures for use of a "What If" checklist. For some hazards, a "What If" checklist will be inadequate and other hazard analysis methods may be used.

Responsibilities

Project Manager

For all phases of the design review, the project manager will be responsible from initiation to completion. That includes initiation of the design review, forming the design review committee, compiling and maintaining the required information, document distribution, setting meeting schedules and agendas, and preparing the final design review report. The project manager will be responsible for coordination and communication with all outside design, engineering, and hazard analysis consultants.

Department Manager

Department managers will see that design reviews are completed for capital expenditure or equipment purchase approvals, and prior to placing equipment or processes in operation, as required under "Installation Review." Signatures of department managers shall not be placed on asset documents until they are certain that all design reviews have been properly completed, and that their findings are addressed.

Design Engineer

Whether an employee or a contractor, the design engineer shall provide to the project manager and to the review committee documentation including:

- detailed equipment design drawings
- equipment installation, operation, preventive maintenance, and test instructions
- details of and documentation for codes and design specifications
- requirements and information needed to establish regulatory permitting and/or registrations

For all of the foregoing, information shall clearly establish that the required consideration has been given to safety, health, and environmental matters.

Safety, Health, and Environmental Professional

Serving as a member of the hazard review committee, the safety, health, and environmental professional will assist in identifying and evaluating hazards in the design process and provide counsel as to their avoidance, elimination, or control. Special training programs for the review committee may be recommended by the safety, health, and environmental professional. Also, consultants may be recommended who would complete hazards analyses, other than for the "What If" system.

Administrative Procedures

In this section, the administrative procedures would be set forth, such as the amount of time prior to submission of a capital expenditure or equipment purchase request is allowed to the Hazard Review Committee for its work, information distribution requirements, advance notice time requirements for Installation Review meetings, procedures to assure that findings of hazards analyses are addressed, and how differences of opinion of hazard review committee members are to be resolved.

Conclusion

Safety professionals are encouraged to venture into safety through design. Opportunities for accomplishment and recognition are great there.

References

1. Kletz, Trevor. *Plant Design for Safety: A User-Friendly Approach.* New York: Hemisphere Publishing Corporation (Taylor Francis), 1991.
2. *Military Standard System Safety Program Requirements* (MIL-STD-882-C). Department of Defense, Washington, D.C., 1993.
3. Haddon, Jr., William J. *Preventive Medicine: The Prevention of Accidents.* Boston: 1966.
4. Haddon, Jr., William J. "On the Escape of Tigers: An Ecological Note," *Technology Review* (May 1970).

Chapter 14

Comments on Hazards and Risks

INTRODUCTION

All risks to which the practice of safety applies derive from hazards. There are no exceptions. Further, the entirety of purpose of those accountable for safety, in fulfilling their societal responsibilities, is to manage their endeavors with respect to hazards so that the risks deriving from those hazards are at an acceptable level.

Thus, safety professionals must have an understanding of hazards and risks and their relationships to fulfill their responsibilities and to be successful in their communications with those to whom counsel is given.

Hazards and *risks* are not synonymous terms, though they are used interchangeably by some authors. Also, in some writings hazards and risks may be equated with exposures and perils. Unfortunately, the literature on hazards and risks and exposures and perils can be baffling. We safety professionals, to be effective communicators, seriously need to establish what we mean when we use the terms *hazards* and *risks.* As John V. Grimaldi and Rollin H. Simonds write in *Safety Management* (1):

> Unless there is common understanding about the meaning of terms, it is clear that there cannot be a universal effort to fulfill the objective they define.

Defining Hazards: Relating Hazards to Risks

An appendix to *Improving Risk Communication* (2) is titled "Risk: A Guide to Controversy." In it, Baruch Fischhoff writes:

> By definition, all risk controversies concern the risks associated with some hazard . . . . the term "hazard" is used to describe any activity or technology that produces risk.

Fischhoff properly relates hazards to risks. Hazards are defined as the potential for harm: Hazards include the characteristics of things and the actions or inactions of people.

Hazards must be considered in the broad context of that definition. Every element within the safety process should serve to avoid, eliminate, or control the aspects of activity or inactivity and the aspects of technology that present a potential for harm and produce risk.

Significance of Hazards within the Safety Process

Whether the concern of a safety professional is occupational safety and health, product safety, environmental affairs, fire protection, transportation safety, or any other safety-related practice, the generic base for that endeavor is hazards. In every one of those fields of endeavor, all activities should be similarly directed to encompass both the possible actions or inactions of people and the characteristics of properties, equipment, machinery, or materials that present the potential for harm or damage.

Further, whatever the safety process—management involvement, safety in the design stages, employee training, hazards communication, incident investigation, use of personal protective equipment, behavior modification, et cetera—its fundamental purpose is to avoid, eliminate, or control hazards.

Hazard Identification

In the definition of hazards given, *potential* is the key word. If a hazard is the potential for harm, and if the hazard is not avoided, eliminated, or controlled, the potential will be realized.

Two considerations are necessary in determining whether a hazard exists. Do the characteristics of the thing or the actions or inactions of people present the potential for harm or damage? And, can people, property, or the environment be harmed or damaged if the potential is realized?

Many methods are available to identify hazards—from historical data, codes and standards, the observations of learned people, and the use of analytical methods such as the more simple "what if" system to the more complex fault tree analysis. And the literature, of which there is a great deal, speaks extensively of those methods.

Defining Exposure

To complete a hazard analysis after a hazard has been identified and evaluated, the exposure must be assessed. An exposure assessment would be a determination of the people, property, and aspects of the environment in a particular setting that could be affected by the realization of the hazard and the extent of harm or damage that could result.

Exposure is not the hazard; nor is it the risk.

Defining Perils

In the insurance-related literature, *peril* is a frequently found term. It may be used synonymously with hazard or risk or exposure. Perils insured against are commonly considered to be fires, explosions, falling aircraft, windstorms, floods, automobile accidents, embezzlements, et cetera. Perils are incidents that occur when hazards are realized; they are events, not hazards, risks, or exposures.

Acquiring Additional Knowledge of Hazards

There are two significant thought processes, one built on the other, of which knowledge is a requirement in professional safety practice. They are Haddon's unwanted energy release concept, extended by him to include exposures to hazardous environments, and the concepts on which the Management Oversight and Risk Tree—MORT—was developed.

I encourage safety professionals to give particular attention to Haddon's concepts and to MORT. Both establish that if there are no unwanted energy releases or exposures to hazardous environments, no hazards-related incidents will occur.

All Work Requires the Expenditure of Energy

In *The Loss Rate Concept In Safety Engineering* (3), R. L. Browning wrote that:

> Work requires the expenditure of energy, in fact, energy is measured by the work it is capable of performing. It follows that the capability to cause—the

> key element in the search for valid loss exposures—will be an inventory of potentially destructive energy.

Browning's statements are thought-provoking. Certainly, being causal-factors-oriented is a fundamental in the practice of safety. And, it seems that "potentially destructive energy," the release of which occurs in hazards-related incidents, is directly related to the definition of hazards previously given.

Obviously, the purpose of people in the occupational setting is to work. All work requires the expenditure of energy. If the energy expended is undesired or excessive in relation to human capabilities, that energy expenditure can be an incident causal factor. Safety professionals cannot do a proper job of giving advice on preventing or mitigating harm or damage without extensive knowledge of causal factors that include unwanted releases of "potentially destructive energy." (Consider: material handling incidents, slips and falls, contact with objects or equipment, et cetera.)

Haddon's Unwanted Energy Release and Hazardous Environment Concept

Dr. William Haddon, the first director of the National Highway Safety Bureau, was the originator of the energy release theory (4, 5). Its concept is that unwanted transfers of energy can be harmful (and wasteful) and that a systematic approach to limiting their possibility should be taken.

Although Haddon stated in his paper "On the Escape of Tigers: An Ecologic Note" (5) that "the concern here is the reduction of damage produced by energy transfer," he also said that "the type of categorization here is similar to those useful for dealing systematically with other environmental problems and their ecology."

Excerpts follow from "On the Escape of Tigers: An Ecologic Note" (5).

> A major class of ecologic phenomena involves the transfer of energy in such ways and amounts, and at such rapid rates, that inanimate or animate structures are damaged.
>
> Several strategies, in one mix or another, are available for reducing the human and economic losses that make this class of phenomena of social concern. In their logical sequence, they are as follows:
>
> - prevent the marshalling of the form of energy
> - reduce the amount of energy marshalled
> - prevent the release of the energy
> - modify the rate or spatial distribution of release of the energy from its source
> - separate, in space or time, the energy being released from that which is susceptible to harm or damage

- separate, by interposing a material barrier, the energy released from that which is susceptible to harm or damage
- modify appropriately the contact surface, subsurface, or basic structure, as in eliminating, rounding, and softening corners, edges, and points with which people can, and therefore sooner or later do, come in contact
- strengthen the structure, living or non-living, that might otherwise be damaged by the energy transfer
- move rapidly in detection and evaluation of damage that has occurred or is occurring, and counter its continuation or extension
- after the emergency period following the damaging energy exchange, stabilize the process

All hazards are not addressed by the unwanted energy release concept. Examples are the potential for asphyxiation from entering a confined space filled with inert gas, or inhalation of asbestos fibers. But all hazards are encompassed within a goal that is to avoid unwanted energy releases and exposures to hazardous environments.

John V. Grimaldi and Rollin H. Simonds in *Safety Management* (1) recognized Haddon's work. Dr. Robert L. Brauer, in *Safety and Health for Engineers* (6), commented on the energy release theory and listed the ten strategies for preventing or minimizing adverse results as outlined by Haddon. In *MORT Safety Assurance Systems* (7), William G. Johnson also listed Haddon's ten energy management strategies and made these comments concerning them:

> Haddon systemized a set of 10 energy management strategies in a progressive order, which can be used in various combinations. . . . Haddon points out that the larger the energy, the earlier in the strategy list should control be sought. This author would add, the larger the energy, the greater the need for redundant, successive strategies and barriers. . . . The systematic review of available strategies and the creation of optimum mixes to reduce harm has not been customary in safety. . . . The application of Haddon's concepts was tested in early trials of MORT.

As a result of the testing and early trials, Haddon's strategy was introduced into the MORT system with an extended hierarchy of preventive measures, numbering thirteen.

Ted Ferry in *Safety and Health Management Planning* (8) wrote about energy transfers and barriers and listed "13 strategies for systematic energy control" as cited by the Department of Energy.

Unwanted releases of energy or exposures to hazardous environments contribute as causal factors for all hazards-related incidents. Haddon's ten energy management strategies were greatly expanded in Chapter 13, "Guidelines: Designing For Safety," under the caption "General Design Requirements: A Thought Process for Hazard Avoidance, Elimination, or Control."

On the Significance of MORT

I also recommend that safety professionals develop an understanding of the concepts on which the Management Oversight and Risk Tree—MORT—is based and the thought process it promotes. In the excerpts that follow, taken from the *Guide to Use of the Management Oversight and Risk Tree* (9), the relationship of MORT to Haddon's original work will be evident:

> This document is the User's Guide for MORT (Management Oversight and Risk Tree), a logic diagram in the form of a "work sheet" that illustrates a long series of interrelated questions. MORT is a comprehensive analytical procedure that provides a disciplined method of determining the systemic causes and contributing factors of accidents. Alternatively, it serves as a tool to evaluate the quality of an existing system.
>
> While similar in many respects to fault tree analysis, MORT is more generalized and presents over 1,500 specific elements of an ideal "universal" management program for optimizing environment, safety and health, and other programs.
>
> MORT conceives the accident occurred when an unwanted energy flow or environmental condition that results in adverse consequences reaches persons and/or objects. MORT combines this concept and others into a functional accident definition as follows: An unwanted transfer of energy or environmental condition because of lack of or inadequate barriers and/or controls, producing injury to persons and/or damage to property or the process.

After an accident occurs, the first step in the MORT process is to consider the adequacy of the amelioration functions, their intent being to limit the consequences of what has occurred. Then MORT users would determine the unwanted energy flow or hazardous environmental condition, whether barriers and controls were less than adequate, and whether vulnerable people or property were exposed.

While MORT is based on the fault tree method of system safety analysis, its logic diagram does not require statistical entries and computations for event probabilities. MORT is presented as an incident investigation methodology and as a basis for safety program evaluation.

Its thought process could also serve well the safety professional who participates in design concept discussions. MORT is soundly based on the unwanted energy release and environmental exposure concept, and its use leads to a good understanding of hazards, exposures, and risks.

A first approach to MORT could be intimidating: its logic diagram presents over 1,500 elements. It can seem daunting, but it should not be so judged. Thousands have successfully completed MORT seminars, and have gained considerably from exposure to its thought process.

I would like to repeat the quotation from Baruch Fischhoff's *Risk: A Guide to Controversy* (2), which appeared earlier in this paper:

> By definition, all risk controversies concern the risks associated with some hazard. . . . the term "hazard" is used to describe any activity or technology that produces risk.

Risks with which safety professionals deal derive from hazards. We need to know more about risks.

Defining Risk and the Role of the Safety Professional with Respect to Risk

Risk is defined as a measure of the probability of a hazards-related incident occurring and the severity of harm or damage that could result.

In producing the measure that becomes a statement of risk, it's necessary that determinations be made of:

- the existence of a hazard or hazards
- the exposure to the hazard
- the frequency of endangerment of what is exposed to the hazard
- the severity of the consequences should the hazard be realized (the extent of harm or damage to people, property, or the environment)
- the probability of the hazard being realized

If the exercise is stopped after the severity of possible consequences is determined, the result is a hazard analysis. To determine risk, the probability of an occurrence must also be estimated. Risk is both a measure of the probability of a hazards-incident occurring and a determination of the consequences of the incident.

Thus, in fulfilling their responsibilities, safety professionals must consider the two distinct aspects of risk:

- avoiding, eliminating, or reducing the *probability* of a hazards-related incident occurring;
- minimizing the *severity* of harm or damage that could result.

As safety practice evolves, the required attention will be given to the avoidance of hazards in the design and redesign processes. As more safety professionals are successful in establishing themselves as consultants in that process, they will become more familiar with concepts of risk and how risk reduction is effectively achieved.

Along the way, questions will arise concerning what the content of professional safety practice ought to be and how it is best applied in giving advice that actually attains a significant reduction of risk.

Some safety practitioners are directed by their managements to apply all of their efforts to obtaining compliance with laws, codes, standards, and regulations. Surely, being in compliance is a laudable goal. Unfortunately, merely being in compliance usually means achieving a minimal level of hazards management. It should not be assumed that actions taken to be in compliance with legal requirements address an organization's principal risks or that doing so, by itself, will attain effective hazards management.

A safety director tells the story of convincing his management to spend $1 million to comply with OSHA standards. Later, he was asked what impact the expenditures would have on the types of employee injuries and illnesses that had been occurring. Injury and illness records were available for thirty years. Not one of the reported incidents was related to the expenditures for OSHA compliance. Yes, risks of employee injury and illness were reduced through the expenditures to comply with OSHA standards, but by how much?

Understanding Management's View of Risk

A successful communication with management personnel on risk is not possible until an understanding has been reached on the meaning of the term as it is to be used in those communications. That's important, because risk is a term with far too many meanings.

On any given day, managements may hear the term in a variety of contexts—in examining a financial venture, in considering the possible increase in the price of a commodity, in receiving a phone call from a stock broker who proposes a risk management program. It's also important that safety professionals appreciate the culture of an organization, the perception of risk a management staff may have, its fears and logic concerning risk, and its tolerance for risk. Managers are risk-takers; so are safety professionals.

On Acceptable Risk

Safety was defined as that state for which the risks are judged to be acceptable. That definition implies a determination of risk, and a judgment of the acceptability of risk. Often, when the term *acceptable risk* is used, the response of the uninformed may be, How dare you suggest that some risk is acceptable?

But consider this: every safety professional who writes a recommendation to eliminate or control a hazard makes a risk acceptance decision. It cannot be presumed that complying with the recommendation achieves zero risk. No thing or activity is risk-free. Also, in the practical world, all risks will not be eliminated. Even when deciding which risks are to be given priority consideration, a risk acceptance decision is made about those risks to be dealt with later, if only for the short term.

A few safety practitioners make too much of participating in risk acceptability decisions. They create an atmosphere of being above that sort of thing, of super-righteousness. Perhaps they believe that if all that they proposed was favorably acted upon, a risk-free environment will be achieved. Safety professionals know that is not possible. If they want to be a part of the management team, they must share in risk acceptance decisions.

Risk Reduction Advice

Safety professionals must acquire knowledge of risk determination concepts to give validity to the proposals they make to reduce risk. Since there are always resource and scheduling and time constraints, the advice given should ideally include:

- risk categorization and priority indications
- possible alternative remediation measures
- expected effectiveness of each alternative in risk reduction
- remediation costs

Risk Management Decisions May Not Be Based on Logic

Although a safety professional may present logically developed data on risks, it should be understood that risk reduction decisions may not be made on that data alone, particularly when dealing with the perceptions the public or employees may have of risk. Richard F. Griffiths wrote in *Dealing with Risk: The Planning, Management and Acceptability of Technology Risk* (10) that:

> The applicability in risk assessment and acceptability is that for low-frequency events the probability estimate is not based on a large number of trials and the public evaluation may well be more conditioned on how bad the outcome might be, with little regard for arguments as to how likely it is.

In this one paragraph, Griffiths introduces two important subjects: that probability estimates used for low-frequency high-consequence incidents may be questionable; and that public concerns may reflect only perceptions of severity of outcomes.

At a symposium on risk sponsored by the National Safety Council and the International Life Sciences Institute, the theme for which was "Regulating Risk: The Science and Politics of Risk," one speaker expressed the view that public risk concerns would best be considered as public outrage. Even though at times it may be believed that the public outrage is illogical, it will often be a significant factor in risk decision-making. All risk reduction measures may not be based entirely on risk assessment logic. And it may be that in dealing with public outrage, employee concerns, or perceived risks, facts will not be convincing.

It will not be unusual when risk decisions are made that the decision process is influenced by elements of fear and dread, and by the perceived risks of employees, the immediate community, a larger public, and management personnel.

Questioning the Validity of Some Quantitative Risk Analyses

Since so many subjective judgments are necessary in applying what are otherwise sound statistical quantification systems, I believe that, except when statistical concepts can be applied to the known, such as the face markings of dice, or when empirical probability evidence has been produced, all quantitative risk analyses are really qualitative risk analyses.

Vernon L. Grose, in *Managing Risk: Systematic Loss Prevention for Executives* (11), appropriately cautioned on several occasions against an overreliance on numbers in determining risk when the numbers may not be sound. A reasoned skepticism would serve safety professionals well concerning the validity of the numbers used in what appear to be precise determinations of both the probability of an incident occurring and an assessment of its consequences.

Both probability and consequence numbers may be greatly overstated or understated. Grose may have exaggerated only slightly when he wrote this:

> Because desire overrode reality, the unfortunate gamesmanship that has evolved supposedly to upgrade reliability and safety has come to be called "numerology," defined in the dictionary as follows: A system of occultism (hidden, secret, or beyond human understanding) involving divinations (the practice of trying to foretell the future by mysterious means) by numbers.

Grose indicates that when there is a demand for numbers and an inventing to fill the need, the numbers eventually appear in print and there is no way to recognize their illegitimacy. Thus a caution—be wary of the numerologists.

Grose is not alone in expressing concern about the validity of risk quantification systems. These quotes come from *Improving Risk Communications* (2), a text prepared under the guidance of the Committee on Risk Perception and Communication as a project of the National Research Council:

> . . . analysts are prone to overlook the ways human errors or deliberate human interventions can effect technological systems. . . . The need to quantify risks as an aid to decision making creates special difficulties because the choice of which numerical measures to use depends on values and not only science.

Emphasis is given here to the phrase "the choice of which numerical measures to use depends on values and not only on science."

I quote again from Griffiths' *Dealing With Risk: The Planning, Management and Acceptability of Technological Risk* (10):

> If one compares the experts' best estimates of the consequences of an accident with the historical record it is often found that the estimates are greatly in excess of the consequences actually manifested.

As a result of many years of experience with estimates of possible property damage losses developed by fire protection engineers, my observation is the opposite of the view expressed by Griffiths. Our findings were that damage estimates were often notably short of what subsequently occurred.

I also offer a caution concerning numerical risk scoring systems. Recently, I was presented with risk data to which a scoring system was to be applied and risk priority ratings given. An incident with a probability of 1 per 100 plant operating years for which devastating results were calculated, including many fatalities, was given a lower priority than a single disabling back injury with a probability of 1 per plant operating year.

A skepticism concerning numerical risk scoring systems is in order. Indeed, be wary of numerical risk systems, and the numerologists.

On Qualitative Hazard Analyses and Risk Assessments

Nevertheless, if safety professionals are to understand risk and include risk determinations in the counsel they give, they will have to deal with incident probabilities and their consequences. Much of their work will require the gathering of opinions of knowledgeable people in developing hazard/exposure scenarios and in making assumptions about occurrence probabilities.

Matrices, of which there are many in the literature, through which risks are measured and categorized on an informed judgment basis, a qualitative basis, have credibility.

On Risk Estimates That Result in a Money Number

Several authors propose that risk be expressed as a precise numerical quantification, in dollars. The computation would derive from a formula through which probability is multiplied by consequence to obtain the expected loss per unit of time or activity. This is a typical formula:

Risk = Probability × Consequence

Where, probability is the event frequency per unit of time or unit of activity, and consequence is loss or cost per event.

Similar formulas abound, intended to produce a finite, numerical quantification of risk in money terms. Their use presumes a knowledge of incident probability and outcomes far more extensive and precise than seems to exist.

The purposes of the definition of risk I have given and the requirements of qualitative risk categorization matrices can be met without attempting to be so precise. This is an example of a risk estimate that does not include a money number.

Assume that the population of the United States is 264 million and that the number of fatalities resulting from automobile usage annually is 44,000. Then, the risk of fatality, *the consequence,* is at a *probability* of 1 per 6000 of population per year.

Conclusion

To be effective, safety professionals must understand hazards, risks, and hazards analysis and risk assessment techniques. In the use of hazard analysis and risk assessment matrices, judgments of incident probability and consequence will often be made on a subjective basis. And such systems can be made to work; they should be considered more art than science.

References

1. Grimaldi, John V., and Rollin H. Simonds. *Safety Management.* Homewood, IL: Irwin, 1989.
2. Fischoff, Baruch. "Risk: A Guide to Controversy." In *Improving Risk Communication.* Washington, D.C.: National Academy Press, 1989.
3. Browning, R. L. *The Loss Rate Concept in Safety Engineering.* New York: Marcel Dekker, Inc., 1980.
4. Haddon, Jr., William J. "The Prevention of Accidents," *Preventive Medicine* Boston: Little, Brown, 1966.

5. Haddon, Jr., William J. "On the Escape of Tigers: An Ecological Note," *Technology Review* (May 1970).
6. Brauer, Roger L. *Safety and Health for Engineers.* New York: Van Nostrand Reinhold, 1990.
7. Johnson, William G. *MORT Safety Assurance Systems.* New York: Marcel Dekker, 1980.
8. Ferry, Ted. *Safety and Health Management Planning.* New York: Van Nostrand Reinhold, 1990.
9. *Guide to Use of the Management Oversight and Risk Tree.* SSDC-103. U.S. Department of Energy, Office of Safety and Quality Assurance, Idaho National Engineering Laboratory, Idaho Falls, Idaho, 1994.
10. Griffiths, Richard F., ed. *Dealing With Risk.* Manchester, UK: Manchester University Press, 1981.
11. Grose, Vernon L. *Managing Risk—Systematic Loss Prevention for Executives.* Englewood Cliffs, NJ: Prentice Hall, 1987.

Chapter 15

Hazard Analysis and Risk Assessment

INTRODUCTION

All hazards do not have equal potential for harm or damage. All hazards-related incidents do not have equal probability of occurrence. Nor will the adverse effects from those incidents be equal.

It is the norm that several hazards will be known to exist at the same time. Giving appropriate advice concerning them requires that:

- hazards are identified, analyzed and categorized as to the possible severity of harm or damage that could result;
- estimates are given of the probability of hazards-related incidents occurring;
- risks are ranked as to their significance and the priority that should be given them;
- costs are determined of alternate proposals to reduce risks;
- expected risk reduction benefits are given for each remediation proposal.

WHY EMPHASIZE HAZARD ANALYSIS AND RISK ASSESSMENT?

Why make so much of analyzing hazards and assessing risks when dealing with several hazards at the same time? This, typically, is what happens: a hazard is identified and it's assumed that injury, illness, or damage can result if the hazard is realized. Or a hazard may be considered a violation of a standard or regulation.

Almost automatically a recommendation addressing the hazard is presented to decision-makers by the safety practitioner.

In that process, little thought would be given to the significance of the hazard in relation to other hazards, to the priority it deserves, or to the possible effectiveness of the measures proposed to achieve risk reduction. That doesn't speak well of real accomplishment or of effective resource utilization.

In the real world, the safety professional has to recognize that:

- some risks are more significant than others;
- resources will always be limited: staffing and money are never adequate to attend to all risks;
- the greatest good to employees, employers, and society is attained if available resources are applied to effectively and economically obtaining risk reduction.

A safety professional implicitly has a responsibility to give the advice that results in:

- efficient use of resources, *on a priority basis,* to avoid, eliminate, or control hazards
- attaining a state for which the risks are judged to be acceptable

Giving Priority to Severity Potential

Since resources are always limited, and since some risks are more significant than others, safety professionals must be capable of distinguishing the more important from the less important. They would then allocate the necessary time to those risks that have the greatest probability of severity of adverse consequences.

We can learn from what is known as Pareto's Law. Pareto observed from analyses of monetary patterns that the significant items in a group will usually constitute a relatively small portion of the total. Those in financial fields often refer to Pareto's findings as the 80–20 rule, with 20 percent of the statistical body representing 80 percent of the total impact.

I have observed over the years, but not through scientific study, that Pareto's principle applies generally to employee injuries and illnesses, fires, auto incidents, product liability incidents, and pollution incidents.

A study made by Employers of Wausau known as "The Vital Few" (1) indicates that: 86 percent of total workers compensation injuries represent only 6 percent of the total cost; 14 percent of total injuries represent 94 percent of total costs; and 2 percent of total injuries (a part of the 14 percent) represent 70 percent of total costs. This study included several hundred thousand cases. It supports Pareto's idea.

Experiences of individual organizations will not fit that distribution precisely, but many safety directors with rather large statistical bases have also observed that the principle applies, generally.

Obviously, the 14 percent of total injuries representing 94 percent of the total costs include the most severe injuries. Assume that resources are limited and that a safety professional shares the responsibility for having resources effectively applied.

When giving hazards management advice, priorities of the employee, the employer, and society in general should be considered. From all viewpoints, it is obvious that hazards presenting the potential for the most severe harm or damage, those for which consequences are most costly, should be given the highest priority.

Whatever the field of endeavor—occupational safety, occupational health, product safety, transportation safety, fire protection engineering, et cetera—Pareto's law applies. And it prompts some interesting questions about whether we spend far too much time on the insignificant.

Only a few authors have proposed that the identification and evaluation of severity potential deserve a special place in the safety process. In *Techniques of Safety Management* (2), Dan Petersen wrote this under the caption "Severity versus Frequency":

> If we study the mass data, we can readily see that the types of accidents resulting in temporary total disabilities are different from the types of accidents resulting in permanent partial disabilities or in permanent total disabilities or fatalities.

Petersen then listed "The Ten Basic Principles of Safety." This is the second principle:

> We can predict that certain circumstances will produce severe injuries. These circumstances can be identified and controlled.

Petersen said that for the following types of situations, severe injuries are fairly predictable: unusual, nonroutine work; nonproduction activities; sources of high energy; certain construction situations; many lifting situations; repetitive motion situations; psychological stress situations; and exposure to toxic chemicals. Petersen got it right.

In *Profitable Risk Control* (3), William W. Allison spoke of the need to address severity potential:

> A basic problem has been the need of a method to enable each facility to identify those severe risks which can result in loss of life, limb, material resources and profitability in that specific facility or in a new operation.

Severity potential needs special attention. Thus, safety professionals must undertake a separate and distinct activity to seek those hazards that present the most severe injury or damage potential so that they can be given priority consideration. To do that effectively, they must be capable of making hazards analyses and risk assessments.

SUBJECTIVITY IN HAZARDS ANALYSES AND RISK ASSESSMENTS

In Chapter 14, "Comments on Hazards and Risks," I discussed the unlikelihood of having accurate numbers representing the probability of incident occurrences and the severity of their consequences in making hazards analyses and risk assessments. I suggested that safety professionals be wary of numerologists, and also of risk scoring systems.

Through experience, I developed an awareness of the folly of making elaborate computations based on subjective assumptions. While making those elaborate computations may create the appearance of being scientific, the activity is unprofessional when the work being done is based on many subjective judgments.

Fire protection engineers on my staff made many computations of property damage and business interruption loss estimates for their clients. A reasonable-worst-case hazard and exposure scenario—a modeling of an event—would be written. It would include assumptions about hazards being realized, where on the property the incident would most likely occur, the value of the facilities and equipment in that area, and the monetary value of the damage to property that could occur. Values of properties used were provided by the client, and they were often inaccurate.

If chemicals were involved and their release could produce a vapor cloud, assumptions were made about the quantity of material released, its characteristics, the point of release, the shape of the vapor cloud that would develop, wind direction and speed, the barometric pressure, et cetera.

Following a long-used concept, the energy potential of the released material was converted into an equivalency of TNT. Computations would be made and circles drawn on maps indicating blast over-pressures in pounds per square inch. Damage levels were reduced as the distance from the blast point increased.

Forgetting the subjectivity of the assumptions originally made, infinite and meticulous computations expressed as dollar loss estimates were made for a variety of blast over-pressure circles. That was done even though it was understood that if one of the assumptions originally made—the variables—was slightly changed, the outcome could be substantively different.

Client personnel would participate in developing hazard and exposure scenarios. They understood the impact of making changes in the variables for which assumptions had been made and that loss estimates at best would be largely subjective.

I made the decision that, for the purposes of producing such property loss estimates, we would make computations to define only the 3 p.s.i. damage circle and assume that the value of the salvage within that circle would equal the value of the damage outside that circle. That greatly limited the calculations and avoided the appearance of being overly scientific. Thus, the property damage loss estimate would be the equivalent of 100 percent of the value within the 3 p.s.i. circle. That became known, initially in jest, as the Manuele Theory of Loss Estimating, and it was applied by many.

Charles A. Pacella, vice president and national coordinator for property services at M & M Protection Consultants, made a comparative study which included a loss estimating system used extensively in the United States, two such systems commonly used in Europe, and the Manuele Theory of Loss Estimating. He wrote that:

> The conclusion from this comparison is that the simplified method requires the least time and effort, has the fewest assumptions, and yields credible, realistic results.

Why should that history be relevant? It indicates that where subjective judgments prevail at the outset, little is gained in making laborious computations as the hazard analysis proceeds.

Making a Hazard Analysis Precedes and Is Necessary to Concluding a Risk Assessment

Results of the loss estimate studies made by fire protection engineers represented the first part of a risk assessment. They began with hazard identification and, presuming that the hazard was realized, considered that which was exposed to determine the possible consequences. *What was produced was a hazard analysis, always assuming the reasonable-worst-case scenario.*

Risk has been defined as a measure of the probability of a hazards-related incident occurring and the severity of harm or damage that could result. A hazard analysis concludes with an estimate of the severity of the consequences of a hazard being realized. A hazard analysis does not require that the probability of an incident occurring be determined.

Estimating the probability of an incident occurring is the additional and following step necessary in concluding a risk assessment.

Developing a Hazard Analysis/Risk Assessment Decision Matrix

While the literature speaks of many hazards analyses and risk assessment techniques, I propose that a safety professional develop a matrix and a thought process that suits client needs.

Several texts include hazard analysis/risk assessment decision matrices. Every matrix I found has been adapted from *Military Standard—System Safety Program Requirements,* known as MIL-STD-882-C (4). Influence of that standard will be obvious in the remainder of this essay.

In considering the development of a matrix and a thought process for its application, an awareness of reality will help:

- Because of staffing and time constraints, it is not possible to know of or to analyze all hazards. Since all hazards are not equal, subjective but learned judgments will be necessary concerning which hazards to study.
- Be aware that some situations defy statistical analysis.
- Keep it simple—you don't have to complicate things.
- Analyzing hazards and assessing risks is an art. It is not a science. Being absolutely certain in determining incident probability or consequence is not possible. A variation of a thought attributed to Descartes applies: if you can't know the truth, you ought to seek the most probable.
- To communicate with decision-makers, terms must have been defined and agreed upon. While identical terms may be used in the several hazard analysis methods described in the literature, those terms may have different meanings.
- Also, in communicating with decision-makers, it would be well to understand their perceptions and tolerance of risk, and appreciate that perceived risks as well as elements of employee and public fear and dread, and client interests, may impact on risk decisions.
- Implementing a logical hazard analysis/risk assessment model is more important than which model is chosen.

Hazard Severity

Hazard severity categories displayed in Table 15.1 are to provide guidance in developing a hazard analysis/risk assessment matrix. Those categories can be applied to the many subsets of the practice of safety.

Adaptation of a variation of Table 15.1 will require developing an understanding of what the terms selected are to mean in an individual operation. Particularly, definitions of the terms *system loss* or *property damage, major* or *minor system* or *environmental damage,* and *severe* and *minor injury* and *occupational illness* are necessary if those terms are to be used.

Hazard Probability

The probability that a hazards-related incident will occur is to be described in probable occurrences per unit of time, events, population, items, or activity. A qualitative probability may be derived from research, analysis, evaluations of the historical

Table 15.1. Hazard Severity Categories

DESCRIPTION	CATEGORY	DEFINITION
Catastrophic	I	Death, permanent disability, system loss, devastating property damage, or environmental damage
Critical	II	Severe injury or occupational illness, or major system, property, or environmental damage
Marginal	III	Minor injury or occupational illness, or minor system, property, or environmental damage
Negligible	IV	No occupational injury or illness, or system, property, or environmental damage

Source: Adapted from MIL-STD-882-C (4).

safety data on similar systems, and from a composite of opinions of knowledgeable people. Supporting rationale for assigning a hazard probability should be documented. Table 15.2 is an example of a qualitative hazard probability ranking.

Achieving Acceptable Risk Levels

Use of a combination of the hazard severity and hazard probability tables given allows a qualitative assessment of risk from which determinations can be made that risks are acceptable or not acceptable, and from which priorities can be set for actions to be taken.

It must be understood that the tables on hazard severity and probability are to serve only as guides and that definitions of terms within them must be refined to suit the needs of a particular operation.

For further guidance, an example is given in Table 15.3 indicating how a combination of the hazard severity and probability charts can be used to develop qualitative risk assessments.

Risk Impact

To discriminate between hazards having the same hazard risk index, a risk impact determination is necessary. An impact determination consists of a review of the effect of an event economically, socially, and politically. (Example: release of a small amount of chemical into a stream may not cause measurable physical damage, but extreme political and social damage could result.)

Table 15.2. Hazard Probability Rankings

PROBABILITY LEVEL	CATEGORY	PARAMETERS
Frequent	A	Likely to occur frequently
Probable	B	Will occur several times
Occasional	C	Likely to occur some time
Remote	D	Unlikely but possible to occur
Improbable	E	So unlikely, it can be assumed occurrence may not be experienced

Source: Adapted from MIL-STD-882-C (4).

Table 15.3 Hazard/Risk Assessment Decision Matrix

	Severity of consequence			
OCCURRENCE PROBABILITY	**CATASTROPHIC-1**	**CRITICAL-2**	**MARGINAL-3**	**NEGLIGIBLE-4**
A Frequent	1-A	2-A	3-A	4-A
B Probable	1-B	2-B	3-B	4-B
C Occasional	1-C	2-C	3-C	4-C
D Remote	1-D	2-D	3-D	4-D
E Improbable	1-E	2-E	3-E	4-E

Hazard risk index	Suggested category
1-A, 1-B, 1-C, 2-A, 2-B, 3-A	Unacceptable: risk reduction action is necessary
1-D, 2-C, 2-D, 3-B, 3-C	Undesirable, written management decision required as to the corrective action to be taken, and when
1-E, 2-E, 3-D, 3-E, 4-A, 4-B	Acceptable, with management review
4-C, 4-D, 4-E	Acceptable, without management review

IT MUST NOT BE ASSUMED THAT FOR THE TWO "ACCEPTABLE" CATEGORIES ACTION IS NOT TO BE TAKEN TO ELIMINATE OR CONTROL THE HAZARDS.

Source: Adapted from MIL-STD-882-C (4).

Residual Risks

No matter what actions are taken to avoid, eliminate, or control hazards, some residual risks will still exist. If the safety process meets its goal, none of the residual risks will be unacceptable.

Conducting a Hazard Analysis/Risk Assessment

In every one of the following steps, it is vital to seek the counsel of experienced personnel who are close to the work or process. Generally, reaching group consensus on the judgments made would be the goal.

1. *Establish the analysis parameters.*
   Select and scope a manageable task, system, or process to be analyzed, and define its interface with other tasks or systems, if appropriate.
2. *Identify the hazards.*
   A frame of thinking should be adopted that gets to the base of causal factors, which is hazards. A determination would be made of the *potential* for harm or damage that arises out of the characteristics of things and the actions or inactions of people. Hazard potential should be kept separate at this point in the thought process, which prompts recognition of severity potential for itself.
3. *Consider the failure modes.*
   Define the possible failure modes that would result in realization of the potentials of hazards.
4. *Describe the exposure.*
   This is still an identification activity; its purpose is to establish and get agreement on the number of people, the particulars of the property, and the aspects of the environment that could be harmed or damaged, and the frequency of their endangerment. It is not easy to do, and help should be solicited from knowledgeable sources. More judgments than one might realize will be made in this process.
5. *Assess the severity of consequences.*
   Learned speculations are to be made on the number of fatalities or injuries or illnesses, on the value of property damaged, and on the extent of environmental damage. Historical data can be of great value as a baseline. On a subjective judgment basis, agreement would be reached on the severity of consequences. In the Hazard Analysis/Risk Assessment Decision Matrix, four severity categories are given: catastrophic, critical, marginal, and negligible. After a category is selected, a hazard analysis will have been completed.

6. *Determine the probability of the hazard being realized.*
   Unless empirical data is available, and that would be a rarity, the process of selecting the probability of an incident occurring will again be subjective. Probability has to be related to intervals of some sort, such as a unit of time or activity, events, units produced, life cycle. Occurrence probability terms in the Hazard Analysis/Risk Assessment Decision Matrix—frequent, probable, occasional, remote, improbable—are commonly used.
7. *Define the risk.*
   Conclude with a statement that addresses both the probability of an incident occurring, and the expected severity of adverse results.
8. *Develop remediation proposals.*
   When required by the risk assessment, alternate proposals for the design and operational changes necessary to achieve an acceptable risk level would be recommended. For each proposal, remediation cost would be determined and an estimate would be given of its effectiveness in achieving risk reduction.

Risk Ranking and Priority Setting

The preceding outline pertains to individual hazard and exposure scenarios. To properly communicate with decision-makers, a risk ranking system should be adopted so that priorities can be established. Since the hazard analysis and risk assessment exercise is subjective, the risk ranking system would also be subjective.

Considering Remediation Costs

Some risk assessment systems include in their formulae a factor for the cost of reducing risk. In those systems, hazards-related incidents with low probability of occurrence and high severity of consequence, classed as catastrophic or critical, may be given a subordinate ranking if the cost to reduce risk is substantial. Categorizations of that sort should be avoided.

Costs to reduce risk surely are a part of management's decision making, but those costs should not be included in the hazard analysis/risk assessment exercise.

Influence of Organizational Culture

Incidents for which consequences are graded as catastrophic and critical must be addressed. Most such incidents would have low occurrence probabilities. In the decision-making for those events, an organization's culture would be determinant—its values, concern for its employees and the public, sense of responsibility, and determination of the risk it can bear.

This sort of question, considered from many views, would typically be asked in the decision process: If it happens, can we stand it? Aspects of an organization's culture cannot be fitted into a scoring system.

Initiating a Hazard Analysis/Risk Assessment System

If a safety professional initiates a more formal procedure for hazard analysis and risk assessment, thought should be given to how those measures might succeed. It may be that what is being proposed requires a significant change in management practice. Thus, the concepts applicable in successfully achieving change would apply. Too much at one time may be disruptive.

A planned effort will be necessary to convince decision-makers that the safety professional's counsel on hazard analysis and risk assessment can be of value. Small steps forward, proving value, are recommended. In preparing for such an endeavor, I suggest that safety professionals:

- develop an awareness of the risk tolerance beliefs held by decision-makers;
- study the approach to be made to decision makers: consider history, risk tolerance, their needs, and how decision makers could be influenced to conclude that what is being proposed is of value in their achieving their goals;
- work with a team of knowledgeable people and obtain agreement on the benefits to be obtained from the use of an additional hazard analysis/risk assessment system, the methodologies to be used, and the meanings of terms in the Hazard Analysis/Risk Assessment Decision Matrix developed;
- determine which risks deserve priority consideration;
- select one or two higher-category hazard/risk situations;
- follow the outline under "Conducting a Hazard Analysis/Risk Assessment";
- continue to try, assess, modify, and try again.

Being Attentive to OSHA's Hazard Analysis Requirements

Estimating incident probability will not be necessary in fulfilling the hazard analysis requirements of OSHA's *Rule for Process Safety Management of Highly Hazardous Chemicals* (5). It has been said that the standard could apply to as many as 50,000 places of employment. A great many of those locations are not within chemical companies.

The hazards analyses required by this OSHA standard deserve a review by safety professionals, even though they may not be immediately affected by the rule. The standard requires that:

> The employer shall perform an initial hazard analysis (hazard evaluation) on processes covered by this standard. The process hazard analysis shall be appropriate to the complexity of the process and shall identify, evaluate, and control the hazards involved in the process.

> The employer shall use one or more of the following methodologies that are appropriate to determine and evaluate the hazards of the process being analyzed. . . . What-If; Checklist; What-If/Checklist; Hazard and Operability Study (HAZOP); Failure Modes and Effect Analysis (FMEA); Fault Tree Analysis; or an appropriate equivalent methodology.
>
> The hazard analysis shall address: The hazards of the process; The identification of any previous incident which had a likely potential for catastrophic consequences in the workplace; Engineering and administrative controls applicable to the hazards and their interrelationships; . . . Consequences of failure of engineering and administrative controls; Facility citing; Human factors; and, A qualitative evaluation of a range of the possible safety and health effects of failure of controls on employees in the workplace.

Under the requirements for "Pre-startup safety review," the employer is required to provide, for new facilities and for significant modifications, a process hazard analysis—among other things.

In no place in the standard is there a mention of occurrence probability. This appears in the preamble to the standard:

> OSHA has modified the paragraph [Note: the paragraph on consequence analysis] to indicate that it did not intend employers to conduct probabilistic risk assessments to satisfy the requirement to perform a consequence analysis.

Yet, safety professionals will have to consider incident probability in ranking risks and in giving advice to decision-makers.

OSHA's *Rule for Process Safety Management of Highly Hazardous Chemicals* has required that many safety professionals, some of whom are not in chemical companies, become involved in hazards analyses. Also, some of the features of this rule should be considered as precursors of things to come, as discussed in Chapter 18, "Anticipating OSHA's General Industry Safety and Health Program Standard, and a Safety Management Audit System."

Resources on Hazard Analysis and Risk Assessment

Many hazard analysis and risk assessment techniques have been developed. In addition to those cited from OSHA's Rule for Process Safety Management of Highly Hazardous Chemicals, these are some of the methodologies mentioned in the literature: Preliminary Hazard Analysis; Gross Hazard Analysis; Hazard Criticality Ranking; Catastrophe Analysis; Energy Transfer Analysis; Human Factors Review; the Hazard Totem Pole; and Double Failure Analysis. There are other hazard analysis systems. P. L. Clemens discusses twenty-five such systems in the article cited in the following listing.

For references on hazard analysis and risk assessment, these works are suggested:

- "A Compendium of Hazard Identification and Evaluation Techniques for System Safety Application" (6) by P. L. Clemens
- *System Safety Engineering and Management* (7) by Harold E. Rowland and Brian Moriarty
- *System Safety 2000* (8) by Joe Stephenson
- *System Safety Analysis Handbook* (9), Richard Stephan and Warner W. Talso, editors
- *MORT Safety Assurance Systems* (10) by William G. Johnson
- *Safety and Health for Engineers* (11) by Roger L. Brauer
- *Managing Risk: Systematic Loss Prevention for Executives* (12) by Vernon L. Grose
- *Military Standard—System Safety Program Requirements* (MIL-STD-882C) (4)

CONCLUSION

Professional safety practice requires that hazards be analyzed, that the risks deriving from those hazards be assessed, and that a risk ranking system be utilized. Also, it must be understood that hazard analysis is the first step in the safety process, and that the quality of all other safety initiatives follows the quality of hazards analyses.

REFERENCES

1. "Pareto's Law and the 'Vital Few.'" Employers of Wausau. Wausau, WI.
2. Petersen, Dan. *Techniques of Safety Management.* Goshen, NY: Aloray Inc., 1989.
3. Allison, William W. *Profitable Risk Control.* Des Plaines, IL: American Society of Safety Engineers, 1986.
4. *Military Standard—System Safety Program Requirements* (MIL-STD-882-C). Department of Defense, Washington, D.C., 1993.
5. *Process Safety Management of Highly Hazardous Chemicals.* OSHA Standard at 1910.119, February 1992.
6. Clemens, P. L. "A Compendium Of Hazard Identification & Evaluation Techniques For System Safety Applications," *Hazard Prevention* (March/April 1982).
7. Rowland, Harold E., and Brian Moriarty. *System Safety Engineering and Management.* New York: John Wiley, 1991.

8. Stephenson, Joe. *System Safety 2000.* New York: Van Nostrand Reinhold, 1991.
9. Stephan, Richard, and Warner W. Talso, eds. *System Safety Analysis Handbook.* Albuquerque, NM: System Safety Society, New Mexico Chapter, 1993.
10. Johnson, William G. *MORT Safety Assurance Systems.* New York: Marcel Dekker, 1980.
11. Brauer, Roger L. *Safety and Health for Engineers.* New York: Van Nostrand Reinhold, 1990.
12. Grose, Vernon L. *Managing Risk: Systematic Loss Prevention for Executives.* Englewood Cliffs, NJ: Prentice-Hall, 1987.

Chapter 16

On Quality Management and the Practice of Safety

INTRODUCTION

There is a remarkable kinship between the principles of quality management and the principles for the practice of safety. Safety professionals involved in soundly based quality management initiatives have opportunities for professional growth and recognition beyond the usual, since that participation allows them to assist in solving problems that impact broadly on quality, productivity, and cost efficiency, as well as safety.

Safety professionals who have responsibilities in quality management have become aware that the processes to be improved in achieving superior quality are the same processes out of which injuries, illnesses, property damage, and environmental incidents occur. And, having had to acquire a new body of knowledge, they speak an additional language.

Surely, American industry has been on a drive to attain recognition for the quality of its products and services. As an example, receiving The Malcolm Baldrige National Quality Award, given by the United States Department of Commerce, has become a mark of prestige. In 1996, over 145,000 copies of the criteria for the award were requested from the Department of Commerce.

To establish the similarities between the principles of quality management and the principles for the practice of safety, I will:

- explore the theoretical ideal for quality and safety;
- review the Criteria for the Malcolm Baldrige National Quality Award;
- comment on the work of Deming, Crosby, Juran, Gryna, Winn, and Brown, Hitchcock, and Willard;

- give emphasis to the Six Sigma quality management program at Motorola;
- establish that quality management and safety initiatives succeed or fail for the same reasons; and
- reflect on my own experience.

On the Theoretical Ideal

A statement in *Why TQM Fails and What To Do About It* (1) by Graham Mark Brown, Darcy E. Hitchcock, and Marsha L. Willard provides a basis for review to determine how near operations are to achieving the theoretical ideal for quality or safety. If the following quotation is read with "safety" replacing "quality," where safety has been inserted in parenthesis, the substance of the statement remains sound:

> When TQM (safety) is seamlessly integrated into the way an organization operates on a daily basis, quality (safety) becomes not a separate activity for committees and teams but the way every employee performs his or her job responsibilities.

Some organizations, according to Brown et al., maintain that "they don't have a total quality management program; rather, they have a culture change initiative." Also, Brown, et al., say "Others insist that TQM is a macrochange strategy. The intent of these companies is to integrate TQM so completely into the organization that it is virtually indistinguishable."

The theoretical ideal is reached when quality or safety is seamlessly integrated into an organization's culture and the way it operates on a daily basis. When that seamless integration is attained, separately identified quality or safety programs would not be needed since all actions required would be blended into daily operations.

Relationship of the Criteria for the Malcolm Baldrige National Quality Award to Requisites for Effective Safety Practice

As the following quality management outline is reviewed, the requisites for effective safety practice should be kept in mind, as the similarities are striking. This outline is in the 1996 Criteria for the Malcolm Baldrige National Quality Award (2), a publication issued by the U.S. Department of Commerce. The Baldrige Award program is administered by the American Society for Quality Control.

1996 Award Criteria—Item Listing

1.0 Leadership
   1.1 Senior Executive Leadership
   1.2 Leadership System and Organization
   1.3 Public Responsibility and Corporate Citizenship
2.0 Information and Analysis
   2.1 Management of Information and Data
   2.2 Competitive Comparisons and Benchmarking
   2.3 Analysis and Uses of Company-Level Data
3.0 Strategic Quality Planning
   3.1 Strategy Development Process
   3.2 Strategy Deployment
4.0 Human Resource Development and Management
   4.1 Human Resource Planning and Evaluation
   4.2 High Performance Work Systems
   4.3 Employee Education, Training, and Development
   4.4 Employee Well-Being and Satisfaction
5.0 Process Management
   5.1 Design and Introduction of Products and Services
   5.2 Process Management: Product and Service Production
   5.3 Process Management: Support Services
   5.4 Management of Supplier Performance
6.0 Business Results
   6.1 Product and Service Quality Results
   6.2 Company Operational and Financial Results
   6.3 Human Resources Results
   6.4 Supplier Performance Results
7.0 Consumer Focus and Satisfaction
   7.1 Customer and Market Knowledge
   7.2 Customer Relationship Management
   7.3 Customer Satisfaction Determination
   7.4 Customer Satisfaction Results

To a large extent, the management processes in which the safety professional has an interest are identical with those of interest to personnel responsible for quality. Whether or not an organization is seeking the Baldrige Award, if the purposes of its quality management endeavors are similar to those set forth in the preceding outline, there is ample opportunity for involvement by safety professionals.

Though the term *customer* does not appear in our literature, safety professionals should understand that they have two categories of customers—customers within operations, and customers external to operations who might be affected by them.

Excerpts follow from the Baldrige Award Criteria for only a few of its categories. In substance, they represent ideals for safety, health, and environmental management.

1.1 Senior Executive Leadership
Describe senior executives' leadership and personal involvement in setting directions and in developing and maintaining an effective performance-oriented leadership system.

1.2 Leadership System and Organization
Describe how the company's customer focus and performance expectations are integrated into the company's leadership system, management, and organization.

1.3 Public Responsibility and Corporate Citizenship
Describe how the company addresses its responsibilities to the public in its performance management practices. Describe also how the company leads and contributes as a corporate citizen in its key communities.

4.2 High Performance Work Systems
Describe how the company's work and job design and compensation and recognition approaches enable and encourage all employees to contribute effectively to achieving high performance objectives.

5.1 Design and Introduction of Products and Services
Describe how new and/or modified products and services are designed and introduced and how key production/delivery processes are designed to meet key product and service quality requirements, company operational performance requirements, and market requirements.

[The following two comments are among several that are a part of this criteria.

Under "Areas to Address": how product, service, and production/delivery process designs are reviewed and/or tested in detail to ensure trouble-free and rapid introduction.

Under "Notes": Applicant's response should reflect the key requirements for their products and services. Factors that might be considered in design include: health; safety; long-term performance; environmental impact; "green" manufacturing; measurement capability; process capability; manufacturability; maintainability; supplier capability; and documentation.]

6.3 Human Resources Results
Summarize human resource results, including employee development and indicators of employee well-being and satisfaction.

If what is expected in an organization in quality management resembles the Baldrige Award requirements, it would seem prudent for safety professionals to be extensively involved. Criteria for the Baldrige Award have a close resemblance to the requisites of good safety management.

Learning from Deming, Juran, and Gryna

Out of the Crisis (3) by W. Edwards Deming is a major reference for this essay. I admit to being a disciple of Deming. (That will be evident.) I will also take excerpts from *Quality Planning and Analysis* (4) by Joseph M. Juran and Frank M. Gryna, which I highly recommend.

Although Deming achieved world renown in quality assurance, it occurred to me as I reviewed his writings that he might regard such a designation as rather narrow in relation to what his work is intended to do. In *Out of the Crisis,* Deming outlined his management theory in a "Condensation of the 14 Points of Management." His position is that his theory must be applied if American industry is to be successful in the world market. Within that theory is his framework for attaining superior quality.

Many authors have quoted Deming's 14 points, sometimes with deviations that are difficult to understand. These are Deming's 14 points, exactly as he wrote them in *Out of the Crisis* (3):

Condensation of the 14 Points of Management

1. Create constancy of purpose toward improvement of product and service, with the aim to become competitive and to stay in business, and to provide jobs.
2. Adopt the new philosophy. We are in a new economic age. Western management must awaken to the challenge, must learn their responsibilities, and take on leadership for change.
3. Cease dependence on inspection to achieve quality. Eliminate the need for inspection on a mass basis by building quality into the product in the first place.
4. End the practice of awarding business on the basis of price tag. Instead, minimize total cost. Move toward a single supplier for any one item, on a long-term relationship of loyalty and trust.
5. Improve constantly and forever the system of production and service, to improve quality and productivity, and thus constantly decrease costs.
6. Institute training on the job.
7. Institute leadership. The aim of supervision should be to help people and machines and gadgets to do a better job. Supervision of management is in need of overhaul, as well as supervision of production workers.
8. Drive out fear, so that everyone may work effectively for the company.
9. Break down barriers between departments. People in research, design, sales, and production must work as a team, to foresee problems of production and in use that may be encountered with the product or service.

10. Eliminate slogans, exhortations, and targets for the work force asking for zero defects and new levels of productivity. Such exhortations only create adversarial relationships, as the bulk of the causes of low quality and low productivity belong to the system and thus lie beyond the power of the work force.
11a. Eliminate work standards (quotas) on the factory floor. Substitute leadership.
b. Eliminate management by objective. Eliminate management by numbers, numerical goals. Substitute leadership.
12a. Remove barriers that rob the hourly worker of his right to pride of workmanship. The responsibility of supervisors must be changed from sheer numbers to quality.
b. Remove barriers that rob people in management and in engineering of their right to pride of workmanship. This means, *inter alia,* abolishment of the annual or merit rating and of management by objective.
13. Institute a vigorous program of education and self-improvement.
14. Put everybody in the company to work to accomplish the transformation. The transformation is everybody's job.

Some companies have built their quality improvement programs on much of what Deming has proposed. But I do not know of an organization that has adopted all of Deming's 14 points. As Rafael Aguayo wrote in *Dr. Deming—The American Who Taught The Japanese About Quality* (5):

> The management lessons of Deming are in direct opposition to what is currently taught in most business schools and advocated by management consultants and business writers.

Deming would do away with all performance reviews, in any form, and performance incentives. In two places in his 14 points, he records his opposition to management by objectives.

Nevertheless, a great deal of what Deming outlines as management theory is directly related to the principles and practices to be applied in attaining successful hazards management. From *Out of the Crisis* (3), *Dr. Deming—The American Who Taught The Japanese About Quality* (5), and *The Deming Management Method* (6) by Mary Walton, I have selected particularly relevant points that associate closely with the practice of safety.

Wherever the word "quality" appears in the following summary, "safety" can be comfortably substituted.

- Quality begins in the boardroom.
- Significant improvement in quality requires a change in the corporate culture.

- A long-term commitment by management and knowledge of what actions must be taken are necessary to measurably improve quality.
- Management support alone will not be sufficient: personal management action and leadership are required. Management obligations cannot be delegated.
- Only management can initiate improvement in quality and productivity. Workers are helpless to change the systems in which they work. On their own, they can achieve very little.
- Everybody has customers. If a person is not aware of who the customers are, that person does not understand the job.
- Quality must be built in at the design stage, where teamwork is fundamental. When processes are in place, revisions are costly.
- Quality is achieved from a continuing improvement in processes.
- Quality comes not from inspection but from improvement of the process. The old way: inspect bad quality out. The new way: build good quality in.
- Inspection to improve quality is performed too late in the system. Mass inspection is ineffective and costly. (There are exceptions to this dictum—e.g., bank statements.)
- Fundamental changes in the system can significantly reduce variations, not adjustments in an existing system.
- Managers have a tendency to put responsibility on workers that is beyond their control. People work in the system created by management, and which only management can change.
- Employees are not a part of the decision making for subjects such as plant layout, lighting, heating, ventilation, product design, process selection and design, determination of work methods, equipment and materials purchasing. Why should they be held responsible for quality?
- Decisions to improve a system should relate to statistical knowledge and thinking. Basing decisions on timely and accurate data is critical. Intelligent decisions can be made only when the data is accurate and properly interpreted. Relying only on data, though, can lead to difficulty.
- Meeting established specifications ensures maintaining the present status. It does not result in improvement.
- Understanding the distinction between a stable system and an unstable system is essential. A statistical chart, with points properly charted, indicates whether or not a system is stable.

Much of what Deming proposes represents a set of principles that a safety professional can easily support. They emphasize: a culture change being necessary to achieve significant improvement in quality; building quality into systems in the beginning; a constant seeking of error reduction and reducing cycle time; a continuing improvement of processes and systems; placing responsibility for the greatest part of what can be done to attain improvement on management; not expecting from employees what they cannot do; and a proper use of statistics.

Crosby: A Different Approach

Philip B. Crosby sets forth a markedly different concept in *Quality is Free* (7). His program is prominently used and some safety professionals say that its adoption has done some good, but I wonder about its staying power. I have difficulty with it, as I do with hazards management initiatives that do not recognize the importance of design and engineering decisions; do not emphasize the management aspects of continuing process improvement; and expect more of employees than they can accomplish.

While there are references to design in *Quality is Free,* the book's argument requires a stretch to conclude that product and process design and management direction of continuous process improvement are important. Crosby presents a program outline that is similar to many safety program outlines. It expects more than can be achieved. It has several rah-rah aspects, with pledges and commitments from employees.

Eventually, the program focuses on Zero Defects, a theme not in vogue now in very many places. The concept supporting the Zero Defects program concerns me. Crosby wrote this (7):

> The Zero Defects concept is based on the fact that mistakes are caused by two things: lack of knowledge and lack of attitude. Lack of knowledge can be measured and attached by tried and true means. But lack of attention is a state of mind. It is an attitude problem that must be changed by the individual.

These premises focus on the employee, on the employee's lack of knowledge and lack of attitude—not on the system put in place by management. Though the Crosby program commences with management commitment, senior management is not very prominent thereafter. It is also my impression that *Quality is Free* expects too much of a quality assurance group and that its emphasis may be more on being in conformance than on reducing conformance ranges through improved product and process design. This is the Crosby program outline (7).

Quality Improvement Through Defect Prevention

Step	Purpose
1. Management Commitment	To make it clear where management stands on quality
2. The Quality Improvement Team	To run the quality improvement program
3. Quality Measurement	To provide a display of current and potential nonconformance problems in a manner that permits objective evaluation and corrective action

4. The Cost of Quality	To define the ingredients of the cost of quality, and explain its use as a management tool
5. Quality Awareness	To provide a method of raising the personal concern felt by all personnel in the company toward the conformance of the product or service and the quality reputation of the company
6. Corrective Action	To provide a systematic method of resolving forever the problems that are identified through previous action steps
7. Zero Defects Planning	To examine the various activities that must be conducted in preparation for formally launching the Zero Defects program
8. Supervisor Training	To determine the type of training that supervisors need to actively carry out their part of the quality improvement program
9. ZD Day [Zero Defects Day]	To create an event that will let all employees realize, through a personal experience, that there has been a change
10. Goal Setting	To turn pledges and commitments into action by encouraging improvement goals for themselves and their groups
11. Error-Cause Removal	To give the individual employee a method of communicating to management the situations that make it difficult for the employee to meet the pledge to improve
12. Recognition	To appreciate those who participate
13. Quality Councils	To bring together the professional quality people for planned communication on a regular basis
14. Do It Over Again	To emphasize that the quality improvement program never ends

Safety professionals could have an interest in certain aspects of a quality improvement program based on Crosby and would want to be active participants in them. But they should be cautious of the emphasis placed on expecting improvements to be initiated by employees, and of the absence of emphasis on management's responsibility for the system put in place, and for its continuous improvement.

Six Sigma at Motorola

A somewhat different approach to quality management from Crosby's has been taken at Motorola through its Six Sigma Program. Motorola was the 1988 winner of the Malcolm Baldrige National Quality Award. The literature on Six Sigma states that defects in the manufacturing process are caused by insufficient manufacturing process control, design margin, or bad material. This places emphasis on product design and on the design and control of the manufacturing process.

What is Six Sigma? It presents an outstanding quality assurance standard. Envision a normal distribution chart, a bell curve. Variability from the mean, the center point of the distribution, is measured in units called Sigma, which is defined as the standard deviation. At plus or minus three Sigma, three standard deviations, 99.7 percent of a population would be included. In a manufacturing process using a three Sigma standard, approximately 2,700 parts per million could be defective.

If a quality performance standard of Six Sigma is achieved, at plus or minus six standard deviations from the mean, it can be expected that no more than 3.4 parts per million will be defective. Four Sigma capability is ten times better than three Sigma capability; five Sigma is thirty times better than four Sigma; and six Sigma is seventy times better than five Sigma. One can appreciate the stretch in the quality standard adopted.

Motorola outlines "The Six Steps To Six Sigma," which, paraphrased, are:

- identifying the product or service you create or provide;
- identifying the customer and what the customer considers important;
- identifying your needs to create the product or service and to satisfy the customer;
- defining the process for doing the work;
- mistake-proofing the process;
- measuring and analyzing, thus ensuring continuous process improvement.

When taking a course on Six Sigma at Motorola University, I was much impressed with the strong emphasis on defining and designing the process for doing the work, mistake-proofing the system, and continuous improvement methods (seeking failure-free methods, standardizing procedures, work simplification, work aids, and training).

Motorola has been highly successful in its quality control endeavors. Several of my safety colleagues have said that their companies visited Motorola for benchmarking purposes. Often, they say that it would take several years for their companies to achieve a quality culture that matches what exists there.

Why Quality and Safety Initiatives Fail

Some quality management ventures, predictably, have not been successful. Brown, Hitchcock, and Willard, in *Why TQM Fails and What To Do About It* (1), comment on the reasons for the failures and offer counsel on how to achieve success. Their text will be of interest to many safety professionals. This comes from the preface to their book.

> Much has been written recently about the failures of total quality management (TQM), as if it is just another management fad in decline, following in the grand tradition of quality of work life programs, quality circles, and the like. . . . many TQM efforts have not yielded the expected results [and] there are common causes for those failures.

Some directors of human relations whose counsel I sought when quality management became prominent expressed concern that it would turn out to be a fad, just as had occurred to many other management panacea. Why were they concerned? They knew that the fundamentals were not being put in place for the culture change necessary to attain the laudable goals set for quality.

Comments of those human relations directors on how employees viewed the new quality management undertakings fit closely with these excerpts from Brown et al. (1):

> Seasoned American employees have seen many programs come and go. They are familiar with the distinguishing characteristics of a three-initial name and the accompanying slogans, signs, banners, bumper stickers, pins, and promises.
>
> Employees remember that programs of the past were introduced with similar fanfare, and consequently, they see TQM as yet another program with slogans and banners.

But, Brown et al. also said that "total quality management is even more critical now than it was 10 years ago." Why do initiatives to improve quality management often succeed or fail? For the same reasons that some undertakings to improve the level of safety succeed or fail. These six premises are near absolutes:

1. Major improvements in quality or safety will not be achieved unless a culture change takes place at all levels of employment, unless major changes occur in the system of expected behavior throughout an organization. Unfortunately, a culture change is not easily accomplished. A colleague has speculated that the time required for the culture change necessary to achieve superior quality or safety is inversely related to the age of an organization.
2. If achieving short-term goals is the management measure, the initiative will fail. Management must be committed to long-term goals and accountability measures, and understand that attaining the culture change necessary may take years.
3. Assessing that which is necessary to achieve customer satisfaction, and preference, must be a continuing and never-ending effort.
4. Systems and processes must be designed to attain superior quality and safety if that is the performance level desired.
5. There must be a continuous improvement initiative, a constant seeking of error reduction, the goal of which is to mistake-proof systems and processes.
6. Operating procedures must be exceptionally well-crafted, recognizing the capabilities and limitations of people, so as to give operating personnel a reasonable chance to achieve the desired quality or safety levels.

What results can be expected if quality initiatives, intended to attain uncommon achievements, are not soundly based? These excerpts, from *Out of the Crisis* (3), are about a typical path of frustration. They also apply to safety.

> A program of improvement sets off with enthusiasm, exhortations, revival meetings, posters, pledges. Quality becomes a religion. Quality as measured by results of inspection at final audit shows at first dramatic improvement, better and better by the month. Everyone expects the path of improvement to continue along the dotted line.
>
> Instead, success grinds to a halt. . . . What has happened? The rapid and encouraging improvement seen at first came from removal of special causes, detected by horse sense. All this was fairly simple. But as obvious sources of improvement dried up, the curve of improvement leveled off and became stable at an unacceptable level.
>
> It takes about two years for people to discover that their program that started off with exhortations, posters, pledges, and revival meetings has come to a dead end. Then, they wake up; we've been rooked.

Deming also stated that "with a sound program the curve of improvement of quality and productivity does not level off."

Gary L. Winn, in an article titled "Total Quality? The 'New' Paradigm Seems Out of Reach for Safety Managers" (8), commented on the abandonment of quality programs by some American companies, and then addressed the difficulty of achieving a culture change.

> . . . why would a company which invested heavily in the Deming or similar Japanese-quality methods be willing to call it quits so soon? The answer may be found in a careful reading of Deming and corroborated by satisfied practitioners in the field.
>
> First, Deming warns that the changes necessary to commit a company to the path of total quality are so far reaching that it may take upper management a year to fully understand the implications, much less implement them. To implement a full gamut of these changes, literally every existing practice must change, and everyone must change. These irreversible modifications to tools, habits, systems and styles are cultural changes, and the successful changes are incredibly pervasive.

Emphasizing System and Process Design

Over time, the level of quality or safety achieved will relate directly to the caliber of the initial design of products, processes, and work methods, and their redesign as continuous improvement is sought.

I plead for an acceptance of what Deming wrote about the import of management setting the stage for design decisions. Deming stressed, again and again, that processes must be designed, or redesigned, to achieve superior quality if such performance is desired, and that superior quality cannot be attained otherwise. And the same principle applies to safety.

For examples of what Deming intended concerning his statement that "quality must be built in at the design stage," I will use the third and fifth points in the previously cited "Condensation of the 14 Points of Management" (3). This is the third point.

> Cease dependence on inspection to achieve quality. Eliminate the need for inspection on a mass basis by building quality into the product in the first place.

To do as Deming proposes requires addressing the quality levels to be attained in the design processes. Similarly, Juran and Gryna, in *Quality Planning and Analysis* (4), speak of "Error Proofing the Process," and I quote from them:

> An important element of prevention is the concept of designing the process to be error free through "error proofing" (the Japanese call it pokayoke or bakayoke). A widely used form of error proofing is the design (or redesign) of the machines and tools (the hardware) so as to make human error improbable or even impossible.

Chapters in Juran and Gryna cover, in depth, "Designing for Quality" and "Designing for Quality—Statistical Tools."

This is Deming's fifth premise, in the "Condensation of the 14 Points of Management" (3).

> Improve constantly and forever the system of production and service, to improve quality and productivity, and thus constantly decrease costs.

The following quotations from Deming's explanation of this fifth premise, taken from *Out of the Crisis* (3), are important.

> A theme that appears over and over in this book is that quality must be built in at the design stage. It may be too late, once plans are on their way. We repeat here . . . that the quality desired starts with the intent, which is fixed by management. The intent must be translated into plans, specifications, tests, in an attempt to deliver to the customer the quality intended, all of which are management's responsibility.

For the paragraphs quoted from *Out of the Crisis,* wherever the word "quality" appears, "safety" can replace it with a good fit. Management "fixes" the level of quality or safety desired, and that "fixing" is a derivation of the organization's culture.

Deming makes a strong case for not expecting from employees what they are not capable of doing. Conversely, it would be folly not to empower employees to do what they are capable of doing.

Problems that are in the system can only be corrected by a redesign of the system by management. If system design and work methods design are the problems, the capability of employees to help is principally that of problem identification.

Because of their emphases on design, these paragraphs were selected from the several paragraphs under "Design Quality and Prevention" in the Criteria for the Malcolm Baldrige National Quality Award (2).

> Business management should place strong emphasis on design quality—problem and waste prevention achieved through building quality into production and delivery processes. In general, costs of preventing problems at the design stage are much lower than costs of correcting problems that occur "downstream." Design quality includes the creation of fault-tolerant (robust) or failure-resistant processes and products.
>
> Consistent with the theme of design quality and prevention, improvement needs to emphasize interventions "upstream"—at early stages in processes. This approach yields the maximum overall benefits of improvements and corrections. Such upstream intervention also needs to take into account the company's suppliers.

These two paragraphs apply precisely to safety. I stress the important point made about economy—that the costs of preventing problems "upstream" in the design stage are much lower than correcting problems "downstream." Also, these paragraphs emphasize prevention and continuous improvement—for which interventions upstream are necessary to achieve superior quality and safety levels.

On Statistical Process Controls

Safety professionals involved in quality management have had to become informed on statistical process control methods. I suggest that a broader application of those methods by our profession to measure safety performance would be beneficial.

Deming's principles, or other soundly based quality management concepts, cannot be applied without the use of statistical control methods: cause-and-effect diagrams, control charts, run (trend) charts, Pareto charts, flow charts, check sheets, histograms. Juran and Gryna are very thorough on this subject.

Deming believed that statistical methods should be used as a guide to understanding accidents and to their reduction. This is from *Out of the Crisis* (3) on reducing accident frequency:

> The first step in reduction of the frequency of accidents is to determine whether the cause of an accident belongs to the system or to some specific person or set of conditions. Statistical methods provide the only method of analysis to serve as a guide to the understanding of accidents and to their reduction.

But I offer a caution concerning an excessive reliance on statistical control methods for the practice of safety, as did Deming on the excessive reliance on statistical controls in quality management.

Low-probability incidents resulting in severe consequences would seldom appear on control charts. Thus, the plottings on a chart indicating that a system is in control can be deluding as to the potential for events occurring that could result in severe harm or damage. Dangerous assumptions could be made by those who are falsely comforted by a control chart indicating that a system is in control.

Conclusion

This essay is intended to be a primer on certain quality assurance concepts and to establish that they have a remarkable kinship with the principles for the practice of safety. I believe that involvement by safety professionals in soundly based quality management initiatives presents opportunities for professional growth and recognition, beyond the usual.

References

1. Brown, Graham Mark, Darcy E. Hitchcock, Marsha L. Willard. *Why TQM Fails and What to Do About It.* Burr Ridge, IL: Irwin Professional Publishing, 1994.
2. 1996 Award Criteria, Malcolm Baldrige National Quality Award. United States Department of Commerce, Technology Administration, Gaithersburg, MD.
3. Deming, W. Edwards. *Out of the Crisis.* Cambridge, MA: Center for Advanced Engineering Study, Massachusetts Institute of Technology, 1986.
4. Juran, J. M., and Frank M. Gryna. *Quality Planning and Analysis.* New York: McGraw-Hill, 1983.
5. Aguayo, Rafael. *Dr. Deming—The American Who Taught The Japanese About Quality.* New York: Carol Publishing Group, 1990.
6. Walton, Mary. *The Deming Management Method.* New York: Putnam, 1986.
7. Crosby, Philip B. *Quality is Free.* New York: McGraw-Hill, 1979.
8. Winn, Gary L., Ph.D. "Total Quality? The 'New' Paradigm Seems Out of Reach for Safety Managers," *Occupational Health & Safety* (October 1994).

Chapter 17

On Safety, Health, and Environmental Audits

INTRODUCTION

Most safety, health, and environmental management audits systems, intended to measure the quality of hazards management in place, are deficient in purpose and content. I include in that observation the systems used for the audits I made, the many audit reports I reviewed, and a study of several audit systems.

To begin with, the principal purposes of an audit and what it is to achieve are seldom understood. More specifically, shortcomings exist in our assessments of an organization's culture regarding safety; the reality of management commitment; design and engineering practices; and procedures in place to identify hazards that could be the causal factors for low-probability/high-consequence incidents. Also, we could do a better job of giving advice on prioritizing risks and the actions proposed to reduce risk, and providing counsel on alternative remediation solutions and their costs.

A SELF-REVIEW OUTLINE

To those who make audits or are responsible for their being made, I offer this self-review outline. It asks the questions that I believe should be resolved to make safety audits more effective.

1. What is the appropriate definition of an audit?
2. What is the *principal* purpose of an audit?

3. How is senior management commitment to safety evaluated?
4. Should audits include an examination of the management practices for the design of the workplace, work methods, and environmental systems and controls?
5. Should it be standard practice to:
   a. prioritize risks and the recommendations submitted to reduce risk; and
   b. give counsel on alternative actions for risk reduction, the costs for each alternative, and the risk reduction expected?
6. Should we be able to give assurance to management when an audit is completed that the:
   a. actual quality of safety has been measured;
   b. principal hazards have been identified and assessed; and
   c. principal management actions necessary to reduce risk are being proposed?

DEFINITION OF AN AUDIT

To begin with, I suggest that safety professionals agree on a definition of a safety, health, and environmental audit. Several writers have provided definitions and addressed what information the audit should provide. Their views differ.

In *Safety Auditing* (1), Donald W. Kase and Kay J. Wiese say:

> Safety auditing, as we will develop it in this work, is a structured and detailed approach to reducing and controlling the seriousness of accidents. . . . And, safety auditing is a method whereby management can receive a continuing evaluation of its safety effectiveness.

In *Safety Audits, A Guide for the Chemical Industry* (2), issued by the Chemical Industries Association Limited, it is implied that the major objective of a safety audit is to determine the effectiveness of a company's safety and loss prevention measures. It is also proposed that whatever the objectives, it is important that they be clearly defined.

In an article titled "Measuring the Health of Your Safety Audit System" (3), Robert M. Arnold, Jr. listed seven benefits that a safety audit system should provide. These two are pertinent to this discussion:

> It offers a precise evaluation of an organization's safety performance.
>
> It establishes the organization's capability to forecast the potential for loss-producing events.

According to William E. Tarrants, editor of *The Measurement of Safety Performance* (4) and author of several of its chapters:

> . . . one must recognize that the main function of a measure of safety performance is to describe the safety level within an organization. . . . in effect, then, measures of safety performance must . . . describe where and when to expect trouble and must provide guidelines concerning what we should do about the problem.

Borrowing from these definitions and reflecting on my experience, I give this definition.

> A safety audit is a structured approach to provide a precise evaluation of safety effectiveness, a diagnosis of safety and health problems, a description of where and when to expect trouble, and guidelines concerning what should be done to improve safety effectiveness.

That's a laudable description. A safety audit that met all of these criteria would truly be professional and serve hazards management needs very well. But do our safety audits really do all that? Most audit systems do not fulfill all of the requirements of that definition. Comments will be made here on the shortcomings.

Principal Purpose of an Audit

Kase and Wiese (1) drew the appropriate conclusion when they wrote that:

> Success of a safety auditing program can only be measured in terms of the change it effects on the overall culture of the operation, and the enterprise that it audits.

An organization's culture determines the probability of success of a safety, health, and environmental process. All that takes place or doesn't take place in that process is a direct reflection of that culture.

The paramount goal of an audit is to influence the organization's culture concerning safety, that is, its system of expected behavior that determines the quality of safety to be obtained. A safety audit report is an assessment of the outcomes of the organization's culture. Even if management complied with every recommendation in an audit report, little would be gained, long term, if there was no change in the overall decision making concerning hazards management. In the typical evaluation system, inadequate attention is given to assessing an organization's safety culture.

Assessing Senior Management Commitment to Safety

Senior management commitment to safety is a part of and a reflection of the organization's culture. And it may be that we don't properly measure senior management commitment. Existence of a results-oriented accountability system that impacts on the financial well-being and the promotion potential of the management staff is the principal measure of whether safety is a subject for which management is held accountable and, thereby, the principal measure of senior management commitment to safety.

It is said that management measures what is important to it. That doesn't go far enough. There are at least two classes of measures—measures of form and measures of substance.

Measures of form are just that. Measures may be produced and observed, such as safety audits, incident statistics, or incident costs, and have no bearing on the status of those responsible for operations.

Measures of substance have an impact on the well-being of those responsible for results as a part of an overall accountability system. Management is what management does, which may be different from what management says. I would ask those who make safety audits whether, when commenting in reports on management commitment, they often give credit for something that isn't really there.

On Design and Engineering Implications

Evidence of an organization's culture concerning safety and its management commitment to safety is first demonstrated through its design and engineering practices—through its upstream design decisions. Yet a requirement for a review of design and engineering practices is seldom included in safety audit guides.

System safety professionals have been trying for years to convince safety generalists that a superior level of safety performance cannot be achieved unless the upstream decision making properly addressed hazards and risks. And they are right. In auditing a safety, health, and environmental process, evaluating the upstream addressing of hazards and risks should be next in importance, following senior management commitment.

Joe Stephenson, in *System Safety 2000* (5), made strong comments on the importance of upstream design and engineering decision-making. These two quotes are from his book:

> The safety of an operation is determined long before the people, procedures, and plant and hardware come together at the work site to perform a given task.
>
> Safety professionals, managers, or supervisors who think they can have a significant impact on safety in the workplace by putting on a hard hat and safety shoes and meandering out with a clip-board to make the world safe are really fooling themselves if the upstream processes have not been properly done. Good safety practices must begin as far upstream as possible.

If a safety audit does not result in influencing the upstream design and engineering practices, it can be expected that there will be a continuation of the same sort of decisions represented by the hazards observed during the audit. While individual hazards may be eliminated or controlled as a result of an audit, similar hazards will more than likely be designed into operations if the concepts to be applied in the design process are not changed.

Several chapters in this book speak of the significance of design and engineering within the practice of safety and they can serve as references for safety professionals who want to be more proficient in assessing safety-related design and engineering systems when making audits.

As an additional reference, I suggest a section of *An Environment, Safety, and Health Assurance Program Standard* (6), which was prepared by Sandia Laboratories for the Department of Energy. That standard sets forth specifications for the establishment and execution of Environment, Safety, and Health (ES&H) Programs. There are five major sections in the program: Management Support, Organization, Program, Line Organization ES&H Functions, and Staff ES&H Functions.

A synopsis follows of what appears under Line Organization ES&H Functions. Note that the functions are assigned to line management. Basically, these functions should be considered as elements of all safety, health, and environmental endeavors, and the auditing practices that are to assess their effectiveness.

Planning: The line organization shall include ES&H considerations in the planning of all operations, and demonstrate through documentation that this has been done.

Facility/Project Status Information: Requires listing all facilities, projects, and activities . . . which present significant hazards . . . and classification . . . by type of hazard.

Identification of Hazards and Their Risks: Significant hazards associated with each activity, project, or facility shall be identified and documented. The risk attendant upon each hazard shall be assessed and documented.

Requirements: ES&H requirements shall be correlated with identified hazards.

Risk Factors: Those factors which determine the risk . . . shall be identified and controls established to adequately limit the risk. Control actions shall be documented and reported to management. The controls shall include those specified in the ES&H requirements and shall consist, as a minimum, of those applicable from the following:

- Funding—Budget proposals shall . . . assure that ES&H needs are taken into account.
- Schedules—[are to see that] . . . adequate time is . . . allowed . . . for . . . safe operation.
- Facilities, Materials, and Equipment—The designs and specifications . . . shall be reviewed to determine that ES&H considerations have been addressed, that hazards have been adequately controlled or eliminated, and that adequate facility environmental controls have been included.

Remaining categories in this section include the qualifications and training of people, written operating procedures, review of changes to determine ES&H

effects, maintenance of records, corrective action, and compliance assurance. One feature of the latter pertains to *audits comparing performance to specifications.*

These provisions presume that the culture and the management commitment include upstream design and engineering considerations and knowledge of hazard identification, evaluation and risk assessment methodologies.

Prioritizing Recommendations and Giving Alternate Solutions

There would be no need to prioritize the recommendations for risk reduction included in an audit report if resources were unlimited and all risks could be scheduled for corrective action immediately. But that is never the case. A safety auditor has an obligation to assign to each hazard/risk for which a recommendation is submitted in an audit report an indicator for the probability of incident occurrence and the severity of harm or damage that could result.

Guidance for the development of such indicators is displayed in a Hazard Analysis/Risk Assessment Decision Matrix in Chapter 15, "Hazard Analysis and Risk Assessment."

Management has a right to ask the safety auditor these questions concerning an audit report:

- What are the most significant risks?
- In what order should I approach what you propose?
- Are there alternative risk solutions?
- Will you help me develop costs for each of the alternatives?
- Will you work with me to determine that what we do attains sufficient risk reduction?

The safety auditor should be able to give helpful responses to these questions. Too often safety auditors submit an audit report containing a laundry list of recommendations without priority indications or an offer of assistance. Audit systems fail if they do not show an appreciation of management needs, if they are not looked upon as assisting management in attaining its goals.

Audit Effectiveness

If the audit is made to fulfill the purposes of the definition of an audit given earlier in this essay, it would provide a precise evaluation of safety effectiveness, a diagnosis of safety and health problems, a description of where and when to expect troubles, and guidelines concerning what should be done about the problems.

If all that is done through the audit process, the auditor should be able to say to management with confidence that the actual quality of hazards management has been assessed, and that the principal management actions to reduce risk are being proposed.

Audits are too often considered shallow and ineffective in relation to actual incident experience. Some audit inadequacies are mentioned in this essay. An often heard criticism of safety audits is that incidents resulting in significant injury or damage occurred after the audit was completed, and that the contents of the audit report had little relation to the causal factors for those incidents.

One reason this occurs is that the audit process does not require attempting to identify those obscure hazards that are the causes of seldom occurring low-probability/high-consequence incidents. To do that, a distinctly identified activity designed for that purpose is needed. Perhaps, in making audits, too much time is spent on the less significant and not enough time on identifying the potential for seldom-occurring incidents that can result in severe harm or damage.

Conclusion

In Chapter 20, "Measurement of Safety Performance," it is said that precise measures of safety, health, and environmental performance are difficult to obtain. Nevertheless, audits can be highly effective measures of the quality of hazards management if they are well conceived and well done. I suggest keeping in mind the observation made by Kase and Wiese (1) concerning the purpose of a safety audit:

> Success of a safety auditing program can only be measured in terms of the change it effects on the overall culture of the operation, and the enterprise that it audits.

References

1. Kase, Donald W., and Kay J. Wiese. *Safety Auditing: A Management Tool.* New York: Van Nostrand Reinhold, 1990.
2. *Safety Audits, A Guide for the Chemical Industry.* London: Chemical Industries Association Limited, 1977.
3. Arnold, Jr., Robert M. "Measuring the Health of Your Safety Audit System," *Professional Safety* (April 1992).
4. Tarrants, William E., ed. *The Measurement of Safety Performance.* New York: Garland Publishing, 1980
5. Stephenson, Joe. *System Safety 2000.* New York: Van Nostrand Reinhold, 1991.
6. *An Environment, Safety, Health Assurance Program Standard,* SAND79-1536. Prepared by Sandia Laboratories, Albuquerque, NM, for the U.S. Department of Energy, 1979.

Chapter *18*

Anticipating OSHA's General Industry Safety and Health Program Standard, and a Safety Management Audit System

INTRODUCTION

Within the next few years, OSHA probably will have promulgated a safety and health program standard for general industry. Possibly, OSHA also will have formalized an audit system to be used by its compliance officers to assess the quality of employer safety and health programs.

If OSHA issues a general industry safety and health program standard and adopts a system to assess the quality of safety management, these two achievements could have more effect on workplace safety in America than all of the specific hazard standards OSHA has thus far promulgated.

In June and December of 1996, OSHA held stakeholder sessions during which interested parties could comment on previously issued documents pertaining to a safety and health program standard. For the June meeting, the paper discussed was "Summary of Provisions Under Consideration For OSHA's Proposed Safety and Health Program Standard" (1). For the December meeting, the subject was "OSHA'S Working Draft of a Proposed Safety and Health Program Standard" (2).

Also, opportunities were provided to submit written comments on the documents prior to the stakeholder meetings.

In October of 1995, OSHA issued an internal document titled "The Program Evaluation Profile (PEP)" (3). Since then, OSHA has been testing and evaluating its safety auditing system.

A Bit of History

There is a history supporting the content of OSHA's proposals for a general industry safety and health program standard and its draft of a safety auditing system. That history should be understood by safety professionals who have responsibilities in occupational safety and health.

The document on which the proposed standard and the auditing system are based is OSHA's *Safety and Health Program Management Guidelines* (4), which were published in 1989. Excerpts from the *Guidelines* are included later in this essay.

In the literature on OSHA's proposed standard, it says that "the rule would provide an effective method to ensure that employers are in compliance with their existing obligations under the general duty clause of the OSHAct and OSHA safety and health standards." OSHA's general duty clause (5) requires that:

> Each employer—shall furnish to each of his employees employment and a place of employment which are free from recognized hazards that are causing or are likely to cause death or serious physical harm to his employees [and] shall comply with occupational safety and health standards promulgated under this Act.

Safety professionals employed at OSHA have been aware for many years that the requirements of employers as set forth in the general duty clause cannot be met unless effective safety and health management systems are in place. Eventually, they surmised, a safety and health program standard for general industry would have to be promulgated.

OSHA's proposed standard is based on work done prior to the adoption in 1982 of its Voluntary Protection Program (VPP) and the changes made in the program in 1988. The following excerpts come from the July 12, 1988, *Federal Register* entry (6), through which the latest changes were announced:

> Requirements for VPP participation are based on comprehensive management systems with active employee involvement to prevent and control the potential safety and health hazards of the site. Companies which qualify generally view OSHA standards as a minimum level of safety and health performance and set their own stringent standards where necessary for effective employee protection.
>
> OSHA has long recognized that compliance with its standards cannot of itself accomplish all the goals established by the Act. The standards, no matter how carefully conceived and properly developed, will never cover all unsafe activities and conditions.
>
> The purpose of the Voluntary Protection Program (VPP) is to emphasize the importance of, the improvement of, and recognize excellence in employer-provided, site-specific occupational safety and health programs. These programs are comprised of management systems for preventing or controlling occupational hazards.

In the preceding *Federal Register* excerpts, the term "management systems" appears more than once. That's significant for the following reasons.

Qualifications for VPP participation are an outline for a safety and health program management system. On January 26, 1989, OSHA published *Safety and Health Program Management Guidelines* (4) in the *Federal Register.* In the background data for the *Guidelines,* these statements were made:

> . . . in 1982 OSHA began to approve worksites with exemplary safety and health management programs for participation in the Voluntary Protection Program (VPP). Safety and health practices, procedures, and recordkeeping at participating worksites have been carefully monitored by OSHA. These VPP worksites generally have lost-workday case rates that range from one-fifth to one-third the rates experienced by average worksites.
>
> Based upon the success of the VPP . . . OSHA published a request for comments and information . . . on July 15, 1988 that included possible language for Safety and Health Program Guidelines.
>
> Based on this cumulative evidence that systematic management policies, procedures and practices are fundamental to the reduction of work-related injuries and illnesses and their attendant economic costs, OSHA offers the following guidelines for effective management of worker safety and health protection.

OSHA's *Safety and Health Program Management Guidelines* became the basis for determining whether a location qualified for the VPP program. They outline a sound safety management system that balances design and engineering, operational, and task performance practices. They are performance oriented.

Verbatim excerpts have been taken from the *Guidelines.* Although brief, they maintain the substance of the *Guidelines.*

EXCERPTS FROM OSHA'S SAFETY AND HEALTH PROGRAM MANAGEMENT GUIDELINES

Scope and Application

This guideline applies to all places of employment covered by OSHA, except those covered by 29 CFR 1926, which is the construction safety standard.

Introduction

Effective management of worker safety and health protection is a decisive factor in reducing the extent and severity of work-related injuries and illnesses.

Effective management addresses all . . . hazards whether or not they are regulated by government standards.

OSHA urges all employers to establish and to maintain programs which meet these guidelines in a manner which addresses the specific operations and conditions of their worksites.

The Guidelines—General

Employers are advised and encouraged to institute and maintain in their establishments a program which provides systematic policies, procedures, and practices that are adequate to recognize and protect their employees from occupational safety and health hazards.

An effective program includes procedures for the systematic identification, evaluation, and prevention or control of general workplace hazards, specific job hazards, and potential hazards which may arise from foreseeable conditions.

An effective program looks beyond specific requirements of law to address all hazards.

Written guidance . . . ensure[s] clear communication of policies and priorities and consistent and fair application or rules.

[Next in the *Guidelines* comes item (b), Major Elements, of which there are four. In a following item (c), Recommended Actions are set forth for each of the Major Elements. In these excerpts, the applicable Recommended Actions follow Major Elements directly.]

Major Element 1

Management commitment and employee involvement are complementary. Management commitment provides the motivating force and resources . . . and applies its commitment to safety and health protection with as much vigor as to other organizational purposes.

Employee involvement provides the means through which workers develop and/or express their own commitment to safety and health protection, for themselves and for their fellow workers.

Recommended Actions

State clearly a worksite policy on safe and healthful work and working conditions, so that all personnel with responsibility . . . understand the priority of safety and health protection in relation to other organizational values.

Establish and communicate a clear goal for the safety and health program and objectives for meeting that goal.

Provide visible top management involvement.

Provide for and encourage employee involvement.

Assign and communicate responsibility . . . so that [all] know what performance is expected of them.

Provide adequate authority and resources.

Hold managers, supervisors, and employees accountable.

Review program operations at least annually to evaluate their success . . . so that deficiencies can be identified . . . [and] . . . objectives can be revised.

Major Element 2

Worksite analysis . . . involves a variety of worksite examinations, to identify not only existing hazards but also conditions and operations in which changes might occur to create hazards.

Effective management actively analyzes the work and worksite, to anticipate and prevent harmful occurrences.

Recommended Actions

Conduct comprehensive baseline worksite surveys for safety and health and periodic comprehensive update surveys.

Analyze planned and new facilities, processes, materials, and equipment.

Perform routine job hazard analyses.

Provide for regular site safety and health inspections.

Provide a reliable system for employees to notify management personnel [of perceived hazards] without fear of reprisal.

Provide for investigation of accidents. . . . Analyze injury and illness trends.

Major Element 3

Hazard prevention and control . . . where feasible, hazards are prevented by effective design of the job site or job. Where it is not feasible to eliminate them, they are controlled to prevent unsafe and unhealthful exposure.

Recommended Actions

So that all current and potential hazards, however detected, are corrected or controlled in a timely manner, establish procedures for that purpose, using the following measures.

Engineering techniques where feasible and appropriate.

Procedures for safe work which are understood and followed . . . as a result of training, positive reinforcement, and, if necessary, enforcement through a clearly communicated disciplinary system.

Provision of personal protective equipment.

Administrative controls, such as reducing the duration of exposure.

Provide for facility and equipment maintenance so that hazardous breakdown is prevented.

Plan and prepare for emergencies, and conduct training and drills.

Establish a medical group.

Major Element 4

Safety and health training . . . addresses the safety and health responsibilities of all personnel concerned.

Recommended Actions

Ensure that all employees understand the hazards to which they may be exposed and how to prevent harm to themselves and others from exposure to these hazards.

So that supervisors will carry out their safety and health responsibilities effectively, ensure that they understand those responsibilities and the reasons for them, including:

- Analyzing the work . . . to identify . . . hazards.
- Maintaining physical protection in their work areas.
- Reinforcing employee training on . . . hazards [and] on needed protective measures, through continual performance feedback and, if necessary, through enforcement of safe work practices.

Ensure that managers understand their safety and health responsibilities as described under . . . Management Commitment and Employee Involvement, so that the managers will effectively carry out those responsibilities.

OSHA'S PROPOSED SAFETY AND HEALTH PROGRAM STANDARD

OSHA's working draft of a proposed general industry safety and health program standard includes the following elements:

- Management leadership and employee participation
- Hazard assessment
- Hazard prevention and control
- Information and training
- Evaluation of program effectiveness

It's not surprising that the proposed standard is a direct adaptation of the *Guidelines.* Probably, when all comments made by stakeholders are weighed and a proposed standard is published in the *Federal Register* for comment, it will look very much like the *Guidelines.* My belief is that OSHA cannot do otherwise.

OSHA says that the Voluntary Protection Program is a success, that participants achieve average injury incidence rates much lower than the relative industry average, and that an aspect of their superior accomplishments is their comprehensive safety and health programs, which must meet the requirements of OSHA's *Safety and Health Program Management Guidelines.*

One of the purposes of having a general industry standard, says OSHA, is to provide an effective method for employers to meet their obligations under its general duty clause. Doing so requires, at least, having a safety process in place that satisfies the requirements of the *Safety and Health Program Management Guidelines.* If OSHA proposes a standard that requires less than the intent of the *Guidelines,* the probability that it will be contested as inadequate is close to certainty.

OSHA's Work on a Safety Auditing System

OSHA's internal instruction on "The Program Evaluation Profile (PEP)" (3) is an intriguing document. Its purpose is to establish policies and procedures for PEP, "which is to be used in assessing employer safety and health programs in general industry workplaces."

Evaluations of workplace safety and health programs would be made and documented; results would determine "whether to conduct a focused inspection, and what recommendations should be made to the employer for improvement." Also, findings would be considered "when calculating an employer's good faith penalty adjustment."

PEP, it is said in OSHA's internal bulletin, is based on the *Safety and Health Program Management Guidelines* issued in January 1989. The bulletin also says that "The Program elements in the PEP correspond generally to the major elements in the 1989 Guidelines." There are six such elements:

1. Management Leadership and Employee Participation
2. Workplace Analysis
3. Accident and Record Analysis
4. Hazard Prevention and Control
5. Emergency Response
6. Safety and Health Training

In the *Guidelines,* Accident and Record Analysis is a subset of Worksite Analysis, and Emergency Response is a subset of Hazard Prevention and Control.

For each of the six elements, scorings are to be made against five narrative descriptions of effectiveness. Subsequently, a composite score for the worksite is to be produced. A summary chart (Table 18.1) follows as it appears in the bulletin. It is duplicated here to give an indication of the scope of the PEP program and the possible implications it could have if it is adopted. "Score" in this chart is the composite score.

Table 18.1. Scoring Levels: Proposed PEP Program

Score	Level of safety and health program	Qualifies for focused inspection?	Good faith penalty adjustment
5	Outstanding program	Yes	80%
4	Superior program	Yes	60%
3	Basic program	Yes	40%
2	Development program	No	15%
1	No program or ineffective program	No	0%

Source: OSHA Instruction CPL2.110—The Program Evaluation Profile (PEP), October 2, 1995.

Table 18.2. Workplace Analysis
Survey and Hazard Analysis

Survey and hazard analysis: An effective, proactive safety and health program will seek to identify and analyze all hazards. In large or complex workplaces, components of such analysis are the comprehensive survey and analyses of job hazards and changes in conditions. [Guidelines, (c)(2)(1)]

1 No system or requirement exists for hazard review of planned/changed/new operations. There is no evidence of a comprehensive survey for safety or health hazards or for routine job hazard analysis.

2 Surveys for violations of standards are conducted by knowledgeable person(s), but only in response to accidents or complaints. The employer has identified principal OSHA standards which apply to the worksite.

3 Process, task, and environmental surveys are conducted by knowledgeable person(s) and updated as needed. Current hazard analyses are written (where appropriate) for all high-hazard jobs and processes; analyses are communicated to and understood by affected employees. Hazard analyses are conducted on jobs, tasks, workstations, where injury or illnesses have been recorded.

4 Methodical surveys are conducted periodically and drive appropriate corrective action. Initial surveys are conducted by a qualified professional. Current hazard analyses are documented for all work areas and are communicated and available to all the workforce; knowledgeable persons review all planned/changed/new facilities, processes, materials, or equipment.

5 Regular surveys including documented comprehensive workplace hazard evaluations are conducted by certified safety and health professionals. Corrective action is documented and hazard inventories are updated. Hazard analysis is integrated into the design, development, implementation, and changing of all processes and work practices.

Source: OSHA, Instruction CPL 2.110—The Program Evaluation Profile (PEP), October 2, 1995.

To give an example of how the program evaluation profile is constructed and the narrative descriptions for which scorings are to be made, the accompanying section, under Workplace Analysis (one of three sections under that title), is being duplicated (Table 18.2).

Obviously, OSHA's PEP Program will be subject to considerable testing and evaluation. Its impact can be momentous.

Other Developments Relative to the Guidelines

Other activity at OSHA previous to the issuance of the "Summary Provisions Under Consideration for OSHA's Proposed Safety and Health Program Standard" and "The Program Evaluation Profile (PEP)" support the premise that the *Guidelines* are considered to be a safety and health management model. Those developments should also be of interest to safety professionals.

In 1990, OSHA issued *Ergonomics Program Management Guidelines for Meatpacking Plants* (7). They are built on the same four program elements contained in the *Safety and Health Program Management Guidelines* issued in 1989: Management Commitment and Employee Involvement, Worksite Analysis, Hazard Prevention and Control, and Safety and Health Training.

In February 1992, OSHA issued its final rule for *Process Safety Management of Highly Hazardous Chemicals* (8). This standard was strongly supported in principle by the chemicals industry. It is a management and performance standard. A multidisciplinary approach will be required in its application. These are the major captions for its requirements:

- Employee Participation
- Process Hazard Analysis
- Training
- Pre-startup Safety Review
- Hot Work Permit
- Incident Investigation
- Compliance Safety Audits
- Process Safety Information
- Operating Procedures
- Contractors
- Mechanical Integrity
- Management of Change
- Emergency Plans and Responses
- Trade Secrets

Although some of these captions may be considered unique to chemicals, all of them fit within the four major elements of the previously issued *Guidelines,* with the exception of Trade Secrets. It's of interest that Employee Participation is the first element. This is what OSHA said about it in the background data issued with the promulgation of the standard:

> Employee participation . . . provisions require that employers consult with employees and their representatives on the general development of a process safety management program, as well as on the process hazards analysis.
>
> OSHA believes that employers *must* consult with employees and their representatives on the development and conduct of hazard assessments and consult with employees on the development of chemical accident prevention plans. . . . OSHA is requiring that all process hazard analyses and all other information required to be developed by this standard be available to employees and their representatives.

The standard requires process hazards analyses and pre-startup reviews, and hazards analyses of existing facilities, new facilities, and significant modifications. These provisions fit closely with the Worksite Analysis requirements of the *Guidelines.*

Management of Change means what it says: "The employer shall establish written procedures to manage changes." This part of the standard is meant to ensure hazard identification and analysis in the change process and that all other aspects of the standard—process safety information, training, et cetera—are met in relation to the findings.

In 1991, a draft was prepared at OSHA representing the views of some personnel on what should be contained in an ergonomics standard for general industry. It also duplicated much of the language of the *Guidelines,* and followed its four program elements. Understandably, medical management was given much more emphasis in that paper, because of the nature of ergonomics injuries and illnesses.

CONCLUSION

Astute safety professionals will be observing with great interest the developments at OSHA with respect to its proposed safety and health program standard and its safety audit system, PEP. Their implications are immense.

REFERENCES

1. "Summary of Provisions Under Consideration for OSHA'S Proposed Safety and Health Program Standard." OSHA, May 6, 1996.
2. "OSHA's Working Draft of a Proposed Safety and Health Program Standard." OSHA, November 15, 1996.
3. "The Program Evaluation Profile (PEP)." OSHA Instruction CPL2.110. Directorate of Compliance Programs, October 2, 1995.
4. *OSHA's Safety and Health Program Management Guidelines. Federal Register 54,* January 26, 1989.
5. Occupational Safety and Health Act of 1970, U.S. Department of Labor, Public Law 91-596.
6. OSHA's Voluntary Protection Programs. *Federal Register 53,* July 12, 1988.
7. OSHA's *Ergonomics Program Management Guidelines for Meatpacking Plants.* U.S. Department of Labor, 1990.
8. OSHA's Final Rule for Process Safety Management of Highly Hazardous Chemicals. *Federal Register,* February 24, 1992.

Chapter 19

Successful Safety Management: A Reflection of an Organization's Culture

INTRODUCTION

Since the first edition of this book was published, several transitions have taken place in the applied practice of safety. To expand on what I have observed, I sought input on current practices from safety professionals in five prominent companies whose safety performance has been consistently outstanding. They are very large conglomerates for which a list of all their business categories would fill several pages.

Each company has been through at least one downsizing and staffs are lean. Throughout their operations, more is expected of fewer people. Although their safety achievements are superior, in all five companies executive-led improvement activity is in progress. (As one CEO/chairman said, "If DuPont can do it, so can we." DuPont has, historically, been a company with a safety record against which many others have benchmarked themselves.)

Transitions that have taken place require expansions of what I originally wrote. I will discuss here how the model companies now achieve superior safety performance and set forth the comments I would make to safety professionals on the specific elements in their safety systems.

CULTURE DEFINED, AND ITS SIGNIFICANCE

In all of the model companies, safety is culture-driven. Senior management is personally and visibly involved and holds its employees accountable for results. The senior executive staffs display by what they do that hazards management is a subject to be taken very seriously, a subject considered in performance measurement along with other organizational goals.

What is meant by "culture"? An organization's culture consists of its values, beliefs, legends, rituals, mission, goals, performance measures, and sense of responsibility to its employees, customers, and community, all of which are translated into a system of expected behavior.

An organization's culture determines the level of safety to be obtained. What the board of directors or senior management decides is acceptable for the prevention and control of hazards is a reflection of its culture. Management attains, as a derivation of its culture, the hazards-related incident experience it establishes as tolerable. For personnel in an organization, "tolerable" is their interpretation of what management does.

Chapters of books and articles with titles such as "The Hazard Control Process," "Basic Safety Programming," "Managing Safety Performance," and "Management of Loss Control" have been written. But none of the authors considered the impact of an organization's culture on the safety performance attainable.

Consider this example of a company in which the safety culture is evident and success in safety is outstanding. In a 194-year-old organization in which it is understood that the chief executive officer is the chief safety officer, the OSHA recordable and lost workday case rates are consistently about one-tenth of the all-industry national average. Employees in this company publicly profess that safety is a part of its heritage, safety is good business, and that safety makes the company credible.

This is a brief summary of its worldwide corporate safety, health, and environmental policy:

- We will comply with all laws and regulations in all manufacturing, product development, marketing, and distribution activities.
- We will routinely review our operations to upgrade beyond legal requirements.
- We will ensure each product can be made, used, handled, and disposed of safely.
- We will inform employees and the public about our products and workplace chemicals.
- We will provide leadership to communities to respond to emergencies.

That policy is believed by most employees to be a sincere statement that is applied by management as written. It establishes a proactive rather than a reactive posture and implies anticipatory hazard prevention and control and the allocation of the resources necessary for accomplishment.

This company's emphasis on off-the-job safety is an additional indication of its culture. Off-the-job safety is treated as importantly as on-the-job safety. It is recognized that the factors influencing behavior and beliefs are the same in both environments. Companies that stress off-the-job safety are the exceptions.

In another organization that also achieves exceptionally good results, an Environment, Health, and Safety Committee consisting of five members of the board of directors sets the corporate requirements and provides oversight of results. Three of the committee members, including the chairman, are "outside" directors. Establishing this committee sent a strong and visible message of intent and accountability throughout the company, and had a significant impact on its culture.

In this company's published "Values," this statement appears under the heading "Environment, Health, and Safety."

> We pledge to protect the environment and the health and safety of employees, the users of our products, and the communities in which we operate.

In an annual report, an extension of its "Values" statement reads as follows:

> We will maintain a leadership position in the protection of the environment, the health and safety of our employees, contractors, the users of our products, and the communities in which we operate.

This company has slightly over 50,000 employees. Results as measured by its records regarding the environment and the safety and health of employees are commendable. Agreement on improvement goals that are rarely achieved has been obtained with its managers.

Since effective safety management is so heavily influenced by an organization's culture, prudent safety professionals will determine what values have been established, management's tolerance for risk, what's important to management, what's attainable, and how things get done, thereby learning about and experiencing the organization's culture. Variations of risk tolerance, particularly by industry, are great.

In the model companies, safety professionals are expected to perform so as to be perceived as a part of the management team and as assisting the decision-makers in fulfilling their expectations. If accomplishment is the safety professionals' purpose, then an understanding must be achieved of the priorities of managers at a given time (expansion, contraction, capital expenditure restrictions, staffing constraints) and of the organization's culture and how to work effectively within it.

A principal goal for safety professionals should be to influence the organization's culture as it pertains to safety decision-making. Understandably, this goal may not be reached easily. Effecting a culture change doesn't get done quickly (a supertanker can't make a sharp right turn). An organization will experience the impact of the culture in place for quite some time. Significant cultural improvement or deterioration occurs only in the long term.

Because of rising costs, public embarrassment, or a number of other factors, management may decide that dramatic improvements in safety must be attained in a rather short time. A bit of skepticism is appropriate when that occurs.

Previously, it was said that the chief executive officer in a certain company is the chief safety officer. But, at its locations, managers are also to function as the chief safety officers, and are to be the role models. Managers cannot delegate this responsibility.

Management Commitment, Direction, and Involvement

In the model companies, senior management assumes responsibility for safety and provides the leadership necessary to achieve the superior results expected. Management has ownership of safety as a part of operating responsibility. It's understood that management commitment, direction, and involvement are the sine qua non, the prime requirement for effectiveness in safety. If superior results are desired, there must be a long-term commitment to long-term goals: that's an absolute.

What management does, rather than what management says, defines the actuality of commitment or noncommitment to safety. What management does permeates the thousands of decisions made that create the work environment, set design specifications for facilities and equipment, establish fire protection standards, respond to environmental needs, et cetera. What senior management does is interpreted by the organization as the role model to be followed. It's at the senior management level that measurable goals are established for performance expectations.

Establishing Accountability

Accountability for safety performance in the superior performing companies is clearly established with line management at every level. Safety performance is one of the elements scored in the overall performance measurement system. Favorable or unfavorable results influence salaries, bonuses, and promotion potential.

One of the principal indicators of management commitment to safety is the inclusion of safety performance in the performance review system. Management commitment to safety is questionable if the accountability system does not include safety performance measures that impact financially and on the promotion potential of those responsible for results.

Here are two real-life indicators of the impact on managers of accountability systems in practice:

1. A plant manager, speaking at a conference, said that the first items discussed in his annual performance review were his achievements in relation to previously established goals for employee injuries and illnesses, environmental occurrences, and fires. Meeting or not meeting those goals had a bearing on his salary. He was very much informed about incidents that had occurred and his involvement was readily apparent.

2. A company became displeased with its employee injury, motor vehicle, and product liability incident experience. Its senior executives arranged a visit to a facility of another organization, known to have a superior incident record. When discussions commenced, visitors were surprised that the meeting was run by the manager of the host location. It became obvious that the safety program was the manager's program and that he considered himself to be accountable for it. The facility manager spoke in depth of his personal involvement in capital expenditure considerations for hazards management, of his requirements for the safety and health professional staff, of the system in place through which he maintained accountability, and of his expectations of the staff immediately reporting to him. During the plant tour, which he led, he commented extensively on the specifics of hazard prevention and control measures in the facility, displaying his personal involvement.

Safety Organization and Staff

Fewer safety professionals are employed now in most of the major companies with which I have had associations in the past few years. Nevertheless, the superior performers still maintain a top-quality staff, which is a requisite for the accomplishment level defined by the culture. The safety staff is expected to earn recognition and respect and establish their capabilities, thereby being sought by decision makers for their views. They are a part of management and have ready access to senior executives.

An organization's personnel will "read" the import given to safety by management through its appraisals of the qualifications of the safety staff and their reporting place in the management structure. If the safety director's position is treated as insignificant, management instructs the organization that safety is insignificant. There is no one magic reporting structure for the safety function, except that the senior safety executive is not far from the top in companies where results are superior.

In one such organization, the vice president for safety, health, and the environment reports to the senior vice president for human resources and corporate plans, who reports to the chief executive officer. In another company, the vice president for environmental affairs and safety reports to the executive vice president, who reports to the chairman.

In both of these organizations, safety, health, and environmental affairs have been brought together under a single management. That is the trend—largely influenced by economics and the recognition that the arrangement presents synergy opportunities. An awareness has developed that the basic sciences of safety, health, and the environment overlap considerably and that greater management effectiveness can be attained under a single direction.

Equally important is the need for effective communication among the professionals involved in each hazards-related function. A case can be made for a unified management which includes all hazards-related professionals.

Professional requirements for safety personnel as to education, experience, accomplishment, and executive ability in those organizations whose cultures require superior safety performance have been moved up a few notches in recent years.

Model companies expect their safety personnel to maintain professional competency and provide opportunities to do so. Their safety professionals are expected to be active in safety committees of trade associations and in technical societies. Also, they are encouraged to expand their horizons through additional education; and increase their knowledge of operations so as to better understand and relate to the organization.

Technical Information Systems

In all organizations where safety expectations are high, technical information systems exist to serve as resources on hazard prevention and control. Personnel at all levels come to rely on those resources. The extent of use of the technical information system is a reflection of the effectiveness of the safety, health, and environmental affairs staff.

Communication and Information Systems

Communications on safety by all levels of employment is encouraged as a part of the organizational culture. Management promotes a continuing and open discussion of hazards, incidents, and concerns about risks. At all levels, personnel are informed of the hazards of operations and of what performance is expected concerning them. Progress relative to established goals is published, discussed, and routinely communicated to employees. Two-way, open communications exist throughout the organization. Thus, the knowledge and experience of employees is brought to bear on improving safety.

Design and Engineering

In the model companies, the first outward indications of their culture with respect to safety are demonstrated through the superiority of their design processes and decisions affecting new and altered facilities, equipment, processes, and products.

Their design and engineering specifications are established to go beyond legal requirements and are intended to avoid unacceptable risk.

Where hazards are given the required consideration in the design and engineering processes, a foundation is established that gives good probability to favorable hazards-related incident avoidance. Also, the potentially large expenses of retrofitting are thus avoided.

This subject, until recently, has not been given sufficient attention by safety professionals. Design and engineering practices do not typically appear in outlines of safety management systems. Nor would the subject ordinarily be included in safety audits. Yet design and engineering decisions are primary in determining the eventual risk level, and they are most often made without input from safety professionals. Thus, safety professionals are typically confronted with the workplace, equipment, and products as givens, with thousands of design and engineering decisions affecting safety having been made without their counsel.

As a better understanding has developed of the phenomena of hazards-related incident causation and as ergonomics has emerged to impact greatly on the practice of safety, safety professionals are required to give greater attention to design decisions. It has been a rewarding experience when safety professionals have been sought for their counsel in concept and design decision-making. There is both need and opportunity here for advice-giving by safety professionals.

Hazard Analysis and Risk Assessment

For the occupational setting, including hazards analyses in design concept discussions, design and engineering decisions, and process reviews is more frequently the practice. There is opportunity here for safety professionals to be perceived as providing a consultancy that produces economic benefits, in addition to improving safety. Those benefits, in addition to reduced costs for injuries and illnesses, are measured by the costs avoided by not having to retrofit to remove hazards brought into the workplace, and by improved productivity and cost-efficiency.

Hazards analyses may be completed through mechanisms as simple as checklists, something as detailed as job hazard analyses, or, in complex cases, failure mode and effect analyses. Whatever the mechanism, the goals are that hazards are to be anticipated, identified, and evaluated, and the appropriate avoidance, elimination, or control measures are to be determined and taken so that the risks deriving from the hazards are at an acceptable level.

Management of Change

Although the title of OSHA's standard for *Process Safety Management of Highly Hazardous Chemicals* suggests a limited application, over 50,000 employment locations could be affected by it. At many of those locations, such as public utilities and paper manufacturers, the final products are not chemicals.

OSHA's standard requires that "The employer shall establish written procedures to manage changes. . . ." The principle involved here is important and is having its impact far beyond chemical companies. What is required is that hazards be identified and evaluated when changes are made in design criteria, operations, procedures, and facilities. That's a good thing to do, and I recommend a broad application of the principle. It is a concept that can be applied in all businesses and industries, as the top performers have learned.

Many companies have said to OSHA personnel that complying with the regulation would take considerable time. In effect, they are saying that the requirement is contrary to the typical practice and that a culture change would have to be achieved to put the required procedures in place. Getting that done in any operation extends the knowledge of personnel about hazards and risks, actually reduces risk, and reinforces safety policy.

Ergonomics and Human Factors Engineering

A conclusion drawn from a study made by a major workers compensation insurer was that about 50 percent of reported claims and about 60 percent of their attendant costs had ergonomics implications. Similar data has been frequently published. As this information developed, safety professionals were required to undertake serious introspection concerning the content of their practice and how they spent their time.

Ergonomics has emerged to become a major element in the practice of safety; that is obvious in the model companies. As its significance grew, ergonomics promoted a greater recognition of the impact of workplace design decisions on both risk reduction and productivity. And safety professionals who acquired the additional knowledge and skill required to be proficient in ergonomics found that decision makers had a greater interest in their work because of its productivity implications. Ergonomics and human factors engineering have become synonymous and interchangeable terms. One university gives courses in both ergonomics and human factors engineering; the difference is that ergonomics covers workplace design, and human factors engineering extends the study to include product design.

Purchasing Standards

A few safety professionals are proud of the growing influence they have had in recent years on the purchasing standards in effect in their companies. Working with design and engineering personnel and with senior executives, they have achieved a culture change that results in fewer hazards being brought into their operations when equipment or materials are purchased.

Including safety-related specifications in purchasing standards assists in attaining corporate goals for superior performance. There is great opportunity here for recognition and accomplishment by safety professionals. In a few companies, suppliers of equipment are not to ship the ordered items until they are visited by the purchaser's safety personnel, who sign off on acceptance.

Preventive Maintenance

The quality of maintenance obviously impacts greatly on hazards management. It sends messages to the entire staff, informing them of the reality of a company's intent to keep or not keep physical hazards at a minimum. While maintenance departments also struggle to get things done with slimmer staffs, safety professionals with whom I recently had discussions say that their companies have not experienced the problem of hazardous situations given a low priority by maintenance staffs.

Visit a location where the culture demands good safety practice and immediately, from the appearance of the exterior premises, you will get a "feel" for the quality of maintenance. This isn't necessarily an absolute indicator, but it is almost always true that if the exterior of the premises is shabby, safety maintenance will likely be inadequate. In the best operations, cleanliness is truly a virtue, maintenance schedules are adhered to, and personnel are encouraged to report on and seek elimination of hazards.

Consider this situation for an opposite and real picture: a safety professional is making an audit of the quality of hazards management. The maintenance superintendent displays an elaborate computer-based maintenance program, of which he is very proud. During the plant tour, many hazardous conditions are observed. A supervisor is asked why work orders aren't being sent to the maintenance department to have those conditions corrected. The response is: "We don't do that anymore. Safety work orders are the last priority for the maintenance department." Later it is determined that a great number of safety-related work orders are over six months old, although the maintenance program, on paper, was supposed to prevent that sort of thing from happening.

A negative message is delivered in a situation of this sort. If the staff is to believe that hazards management is to be taken seriously, management must maintain a safe environment and continuously demonstrate its commitment to do so.

Safety Committees

Although many articles questioning the value of safety committees have been written, in entities where superior results are expected and achieved safety committees are made to work. For the superior performers, it is a common practice for the management committee to also serve as a safety committee, with safety as an early item on meeting agendas.

Safety committees exist at several levels in the model companies. Where they are programmed to achieve, they:

- serve as a means of communicating that hazard prevention and control is important within the organization's culture;
- provide opportunity for participation in safety efforts by a large number of employees;
- can be structured to allow greater employee involvement and upward communication;
- are well-organized;
- have clearly understood purposes;
- find that their recommendations are seriously considered and resolved at appropriate management levels.

Where safety committees are effective, they add to the element of trust, from the top down and the bottom up. Surely, if they are not effective, their existence can further the belief that management is not serious when it says it is concerned about safety.

Supervisory Participation and Accountability

Supervisors in the top performers are conveyors of the element of trust between management and operations employees. Participation by supervisors in hazards management directly reflects the perception their superiors have of what the organization's culture expects and what they understand to be the actual performance measures.

Supervisors will do what they perceive to be important to their superiors. If their superiors convey, by what they do, that hazard prevention and control are important, be assured that supervisors will so respond. If supervisors are held accountable for the prevention and control of hazards, success will result.

Expectations of supervisors, by their superiors and by society in general, have unfortunately become complex and difficult to attain—which means that supervisors must have a sound support structure to be successful. That support structure begins with the location manager and the staff immediately subordinate to the manager. It includes depth of training, a good communication system on hazards, up and down, and the resources of qualified safety professionals as consultants.

Training

In companies with superior safety records, training is serious business. Unfortunately, safety training is often much talked and written about but poorly implemented. Senior managers in the model companies are well-trained. It all starts here. All levels of management become aware of the risks of their business and acquire the necessary knowledge of hazards management needs. They cannot be role models and provide the necessary leadership if they are not schooled in how the hazards management job is to be done.

Training takes place in many ways—in formal classroom settings and on the job by demonstration and observation. It is a continuing and never-ending process.

Safety training must be well planned, continuous, and measured for results. Supervisors and employees have to believe that the content of the training program is what management expects them to apply, and that it serves real knowledge and skill requirements.

Employees cannot be expected to follow safe work practices if they have not been instructed in the proper procedures. They need to understand when they begin employment that they have entered an organization that gives high priority to safe performance. It's typical to have a very thorough indoctrination procedure for new employees. As they pass through indoctrination and are assigned to a supervisor, they are able to evaluate the level of safety expected very quickly.

Too much emphasis cannot be given to the importance of the supervisor in employee training, or to the priority given to training in those companies where successes in hazards management are noteworthy. Supervisors, and experienced employees serving as lead persons, are the role models that new employees will follow.

But consider this situation as representative of a reality that is too prevalent. Early during a safety audit, an industrial relations director proudly reviewed with the auditor a marvelous indoctrination and safety training program for new employees. During the audit, an interview was arranged with an employee who had been in the shop for about three months. The intent was to determine what he thought of the indoctrination and safety training program.

His response was "What indoctrination and safety training program?" This employee had bid up to his third job, had never gone through the indoctrination and safety training program, complained that he never saw his supervisor, and didn't know how to get anyone to pay attention to gear box covers that had been removed and not replaced. Situations of that sort define the organization's culture for hazard prevention and control.

Training needs are always in transition, and recent developments require different emphases. Safety professionals interviewed spoke of these situations:

- New technology is continuously developed that may not have been evaluated for safety. Thus, safety professionals are more often engaged in preoperational hazards analyses and job hazard analysis, and the additional training those analyses indicate is necessary.
- It is more common for employees with seniority to be assigned new jobs without adequate training, and that requires particular attention by the safety staff.
- The changing workforce stretches training capacity to its limits.

Overall, safety professionals say that it has become very difficult to keep up with training needs.

Safety training, given as a substitute for other hazard prevention and control actions that should be taken, will be recognized as the deception it is. Several authors have cautioned that employers should not consider safety training as the primary method for preventing workplace incidents. That premise is sound. Rather, the first course of action when considering hazards should be to determine whether they can be eliminated through workplace and work methods redesign. Training will be less effective if known hazards are not corrected; if those hazards continue to exist, the purposes of the training initiative will be questioned.

Employee Involvement

Safety professionals in the better performing companies agree that effective employee involvement builds confidence and trust in the organization, develops more enthusiastic and productive employees, and supports the position that all are working together to achieve understood objectives. Employees must believe that they are responsible for their safety, and they must be provided with the training, tools, and the necessary authority to act.

Given the necessary training and opportunity, employees can make substantive contributions in hazard identification, in proposing solutions to problems, and as participants in applying those solutions. Safety and health initiatives are obviously more effective if employees have "bought into them."

As an example, practitioners in ergonomics tell countless stories of work practice innovations originating from first line employees. Many are easy to apply, inexpensive, and effective, and often result in greater productivity. There is an asset in effective employee involvement that could be better utilized to achieve more effective hazards management.

Control of Occupational Health Hazards

A major emphasis of OSHA since its beginning has been the control of occupational health hazards. These high-performance companies have given the subject priority attention. Surely, keeping occupational health hazards at an acceptable risk level is a must, though expenditures to control health hazards are great.

Environmental Controls

As a matter of good citizenship and because of concerns over costly penalties that might be imposed by the Environmental Protection Agency, avoiding environmental incidents often gets greater senior management attention than other aspects of hazards management. It is common in the best situations for those responsible for environmental affairs to have senior level credentials, and have management's support to achieve goals.

Safe Practice Standards

A safety initiative cannot succeed without soundly established and implemented safe work practices. How well that's done is another reflection of an organization's culture. It is understood in superior performing companies that establishing, communicating, and implementing prescribed work practices are to be taken very seriously at all employment levels. Developing safe practice standards more often includes some form of employee involvement through which their input is sought. These work standards become the substance of training programs and of expectations by supervisors.

Inspection Programs

Well-managed inspection programs will exist at several levels where hazard prevention and control are managed best. They have many purposes; one of the most important is displaying and communicating management's determination that hazardous conditions and practices are to be identified and corrected. They also provide meaningful opportunities for participation by a cross-section of all employment levels. No inspections are more effective than those in which senior executives participate. Correcting observed hazards is a demonstration of the culture. Failure to follow through, of course, gives negative messages.

Incident Reporting and Investigation

For several reasons greater emphasis is being given to incident reporting and investigation in the model companies with which discussions were recently held. What gets done when a hazards-related incident occurs is one of the major influences that determines how the staff "reads" what level of hazards management is really acceptable. Do it poorly, and poor readings are inevitable. But how does it get done well? Management has to be a part of the accountability system for incident investigations. In one company, the plant manager is expected to participate in at least 10 percent of incident investigations. In another world-wide company, the location manager (not the safety director) must report to headquarters within forty-eight hours on any injury resulting in lost workdays.

Far greater use is being made of incident investigation teams. Safety professionals say that the time expended by those teams is a worthwhile investment since the activity communicates management's intent to avoid hazards-related incidents. Over time, large numbers of personnel are involved.

Absolutely, there has to be a documented incident reporting and investigation procedure. But that's not enough. It's recognized in the model companies that specialized training is necessary to achieve sophistication in incident investigation. Incidents don't occur in a given department very often, and those who investigate them have limited investigation experience. Thus, the necessary training must be repeated.

Most importantly, results of investigations and the actions to be taken concerning causal factors are publicized. Quality of incident reporting and investigation required or tolerated is a principal measure of the accountability system, and of the culture of which the accountability system is a part. It's very difficult to achieve effectiveness in other aspects of hazards management if corrective action is not taken concerning the causal factors for the incidents that do occur.

Recording, Analysis, and Use of Incident Data

If the accountability system is to work, there has to be an effective incident information-gathering and analysis system. Those systems are effective in the model companies. Performance reviews that hold management personnel accountable for meeting or not meeting agreed-upon goals rely on these systems. Also, the analytical data produced is vital in determining where hazard prevention and control emphasis needs to be given.

Performance Measurement

Data produced by the incident recording and analysis system is, of course, a principal aspect of performance measurement. But additional proactive measurement and communication systems are used in the model companies, such as those discussed in Chapter 20, "Measurement of Safety Performance."

Because of the emphasis given to them in discussions with safety professionals, two performance measurement systems deserve further comment. Scheduled safety audits are performed in every superior performing company. Through a formal process, they provide management with a determination that these expectations are or are not being met. In that process a systems aura prevails—an aura of plan, do, check, and improve.

And, for performance measurement these companies benchmark with others in similar businesses, formally and informally. They publish and exchange statistics on incident experience. Through their trade associations and professional society meetings and publications, they explore ideas on how safety can be improved.

Medical and First Aid Facilities

At both a corporate and a location level, medical and first aid facilities are superior where a sense of responsibility to employees permeates the culture.

Emergency and Disaster Planning

Those companies dedicated to protecting their employees and community provide the resources necessary to establish and maintain sound emergency and disaster planning. But, with sympathy, it needs to be said that it's very difficult to put in place and maintain activities that are seldom used. Expectations of emergency and disaster plans cannot be fulfilled without regularly testing their ability to deliver. Establishing communications with the community resources is necessary, without which the actions expected when an emergency occurs will not take place. Training and practice requirements are considerable.

Compliance with Government Regulations

While mentioned last in this essay, compliance with government regulations is important at a corporate level and the needed attention to them percolates down through entities that are to maintain a top quality of hazards management. But compliance programs do not determine operating standards. It's common in the best situations for government regulations to be considered basic standards, with actual design and operating requirements exceeding them.

Conclusion

Listings of the elements of successful safety management always commence with management commitment and involvement. One could argue that management commitment and involvement are not elements to be placed on a par with other elements in the listing. Rather they are the foundation, reflection, and extension of the organization's culture from which all hazards management activities derive. Management involvement and commitment are absolutely required.

In entities that have achieved outstanding safety records, all employees *know* that management is held accountable, is involved, and holds subordinates responsible for their results. If incident experience is considered to be unsatisfactory by management, safety professionals should promote, with great tact and diplomacy, the asking of the obvious but difficult questions. Has that experience resulted from an absence of management commitment to hazard prevention and control? Has the adverse experience been programmed into operations, by implication?

It is impossible for superior safety performance to be attained if executive personnel do not display, by their actions, that they intend to achieve it. Management is what management does. What management does establishes the organization's culture. If what management does gives positive impressions, it is more than likely that a safety initiative will succeed.

References

1. Denison, Daniel R. *Corporate Culture and Organizational Effectiveness.* New York: John Wiley & Sons, 1990.
2. Leavitt, Harold J. *Corporate Pathfinders.* Homewood, IL: Dow Jones-Irwin, 1986.
3. Grimaldi, John V., and Rollin H. Simonds. *Safety Management.* Homewood, IL: Irwin, 1989.
4. Browning, R. L. *The Loss Rate Concept in Safety Engineering.* New York: Marcel Dekker, 1980.
5. Brauer, Roger L. *Safety and Health for Engineers.* New York: Van Nostrand Reinhold, 1990.
6. Ferry, Ted. *Safety and Health Management Planning.* New York: Van Nostrand Reinhold, 1990.
7. Petersen, Dan. *Techniques of Safety Management.* Goshen, NY: Aloray, 1989.
8. Gloss, David S., and Miriam Gayle Wardle. *Introduction to Safety Engineering.* New York: John Wiley & Sons, 1984.
9. Johnson, William G. *MORT Safety Assurance Systems.* New York: Marcel Dekker, 1980.
10. *Process Safety Management of Highly Hazardous Chemicals.* OSHA, 29 CFR 1910, February 24, 1992.

Chapter 20

Measurement of Safety Performance

Introduction

Metrics became a frequently used term among safety professionals in the recent past. A renewed interest has developed in having measurement systems that effectively assess occupational safety performance and are universally applicable. Preferably, those measures would not only be historical but also predictive, and serve as a base from which to prioritize future efforts. A significant goal is to have those measures communicate well in terms that managements understand.

This renewed interest in performance measures derives from the increased desire of some environmental, safety, and health professionals to move the profession forward by being able to establish more definitively the value of their work in relation to organizational goals.

In more than one place, money is being spent in attempts to develop predictive measures of safety performance that do not now exist, as well as to determine how to make better use of existing measures. One group is attempting to formulate a predictive method, model, or process to guide companies on the financial impact of safety and health investments on overall business performance. That group concluded that OSHA rates and workers compensation costs are "after the fact" measures or "trailing measures of failure," and thereby inadequate. I believe that actuaries would not agree with that premise. Nor do I.

This renewed interest in measures of safety performance prompted a review of the literature on the subject as a prelude to setting forth in this essay the significance of measurement as a component within the professional practice of safety; the requirements of a sound measurement system; some guiding thoughts; and the reality of what is conceptually valid, usable, predictive, and attainable.

A Literature Review

Computer searches were made of the records of the American Society of Safety Engineers and the National Safety Council for publications relative to performance measurement. Since 1980, no articles have been published in the journal *Professional Safety* that had "performance measures" or "performance measurement" in their titles or abstracts. A further search was made using "effective" and "effectiveness" as the key phrases.

Two articles were found: the June 1981 issue of the journal contained "How Do You Know Your Hazard Control Program Is Effective?" (1), written by Fred A. Manuele; the February 1989 issue included "Using Perception Surveys to Assess Safety System Effectiveness" (2) by Charles W. Bailey and Dan Petersen.

The National Safety Council library had two relative references in its computer bank: a 1980 book titled *The Measurement of Safety Performance* (3), for which William E. Tarrants was the editor as well as the author of some of its chapters; and a paper published in August 1982 titled "One Method for Evaluating Safety Performance in Working Places" (4) by Shigao Hanayas.

Only one other book, besides the Tarrants text, was located that treated safety performance measurement in any depth: a chapter titled "Measurement Tools For Management" in *Loss Control Management* (5) by Frank E. Bird and Robert G. Loftus. Several texts were found that gave the measurement of safety performance cursory treatment.

In 1995, the Chemical Manufacturers Association issued an important publication on "Program Performance Measures." Its introduction indicates that "A group of member company industrial hygienists and safety professionals recently acknowledged a need for additional guidance on voluntary internal company performance measures." The purpose of the group was to "identify additional objective and voluntary performance measures to assess the effectiveness of occupational safety and health programs." (6)

It is significant that users of "Program Performance Measures" are encouraged to develop their own performance measures, tailored to suit specific company needs. That theme will be developed further in this essay.

Some background is provided here concerning publication of *The Measurement of Safety Performance* (3) since it refers to activities that safety professionals should promote again. Tarrants comments on symposia addressing safety performance measurement held in 1966 and 1970 under the sponsorship of the Industrial Division of the National Safety Council. The first symposium was held, as Tarrants says, "for the purpose of studying the current methods and needs for the measurement of safety performance by employers and establishing measurement methods that lead to total accident prevention." The second symposium concentrated on aspects of measurement as a concept, and safety performance evaluation as viewed by persons primarily outside the occupational safety field. Tarrants' book is a composite of the papers presented at those symposia, along with chapters he wrote.

If major endeavors through symposia or any other type of forum to disseminate knowledge about the measurement of safety performance have been held since 1970, their proceedings remain unknown. Renewed interest in having valid performance measures indicates that we would benefit from additional symposia on the subject.

Significance of Performance Measures in the Practice of Safety

If we safety professionals state that our practice is based on sound science, engineering, and management principles, it logically follows that we should be able to provide measures of performance that reflect with some degree of accuracy the outcomes of the hazards-management initiatives we propose. Understanding the validity and shortcomings of our performance measures is an indication of the maturity of the practice of safety as a profession.

Safety professionals must accept that the quality of the hazards-management decisions made are impacted directly by the validity of the information they provide through the performance measurement systems they put in place. Their ability to provide accurate information to be used in the decision-making process is a measure of their effectiveness.

In this paragraph, I paraphrase Tarrants (3): Measurement is a prerequisite for control and prediction. The main function of a measure of safety performance is to reveal the level of safety effectiveness. A second purpose is to provide continuous information concerning the safety state. Measures of safety performance must help prevent, not just record, incidents. They must indicate where hazards-related incidents will likely occur and provide guidelines concerning the appropriate preventive initiatives.

Requirements for an Ideal Safety Performance Measurement

Assume that a safety professional undertook to develop the ideal instrument for measuring safety performance. For references, numerous books and articles are available on the subject of performance measurement. Two safety-related texts that could also serve as references are those by Tarrants, and Bird and Loftus. In this exercise, the safety professional would probably look for a standard against which the validity of the measurement system developed could be assessed. Such a standard follows. I borrowed from several sources in its development. It is generic.

Regardless of what is to be measured, the following traits would be considered in developing, or assessing the value of, a performance measuring system:

Administrative Practicality. One must be able to develop and use the system within typical, practical management limits.

Measurement Criteria Should Be Quantifiable. Criteria selected must be quantifiable with a good degree of accuracy.

Sensitivity. Measures should be sufficiently sensitive to detect changes in process and performance levels.

Reliability. The technique should be capable of producing the same results from successive application to the same situation.

Stability. If a process does not change, the measure of its performance level should remain unchanged.

Validity. Of prime importance is the need for a measure to be valid. This means that it produces information that is representative of what is measured.

Objectivity and Accuracy. Measures used would, desirably, yield precisely objective and error-free results (a state for which science has not yet produced such an instrument).

Efficiency and Understandability. A good measurement technique should be both efficient and understandable. Efficiency requires that the cost of obtaining and using the instrument is consistent with the benefit to be gained. To be understandable, the criterion must be understood by those charged with the responsibility for approving and using it.

No single safety performance measurement system has been designed that meets all of the preceding requirements. And it may not be possible to do so. Nevertheless, to determine their reliability, the measurement systems used should be tested against these characteristics.

Some Guiding Thoughts

An understanding of the nature of hazards and the risks that derive from them is necessary in determining which performance measurement systems can give reasonably accurate assessments of the quality of safety, and the extent to which those systems can be predictive. To help in developing that understanding, this outline of thoughts is offered.

1. The entirety of purpose of those accountable for safety is to manage their endeavors with respect to hazards so that the risks deriving from those hazards are acceptable. *How well those endeavors are managed is what is to be measured.*
2. Safety results are determined by an organization's culture: its values, beliefs, legends, goals, emphases, performance measures, its management

commitment, and sense of responsibility to its employees, customers, and community.

3. A system of expected behavior derives from an organization's culture, which is demonstrated by its policies, standards, and procedures, and its implementation and accountability systems.
4. To the extent that policies, standards, and procedures, or their implementation, are Less Than Adequate, hazards that derive from design and engineering, operations management, and task performance practices will have been integrated into systems and processes over time.
5. Unless radical events take place, an organization's culture changes slowly. It is commonly said that achieving a major culture change could take as long as seven or eight years.
6. Significant accomplishments in avoiding, eliminating, or controlling hazards that derive from aspects of the culture and that are inherent in an operation can be achieved only over considerable time.
7. A hazard is defined as the potential for harm: Hazards include all aspects of technology (characteristics of things) and the actions and inactions of people.
8. All occupational risks of injury or illness derive from hazards. There are no exceptions.
9. Risk is defined as a measure of the probability of a hazards-related incident occurring and the severity of harm or damage that could result.
10. Realizations of the potentials of hazards have various occurrence probabilities and severities of consequences. Definitions of probability and severity can be tailored to suit particular needs. Indicators given in the following disĭplays are for example purposes only.
11. As the sample base, the number of hours worked, increases in size, the historical incident record (assuming consistency and accuracy in record-keeping) has an increasing degree of confidence for incidents that have frequent, probable, occasional, and sometimes remote occurrence probabilities as:
    - a measure of the quality of safety in place
    - a general, but not hazard-specific, predictor of the experience that will develop in the future
12. But no statistical, historical performance measurement system can assess the quality of safety, to include risks for which the probability of occurrence is remote or improbable, and the severity of outcomes is critical or catastrophic, since such events seldom appear in the statistical history.
13. Even for the large organization with significant annual hours worked, in addition to historical data, hazard-specific and qualitative performance measures are also necessary, particularly to identify low-probability/severe-consequence risks.

Table 20.1. Occurrence Probability Indicators
Must Relate to a Specified Unit of Time or Activity

Frequent	Likely to occur frequently
Probable	Will occur several times
Occasional	Likely to occur sometime
Remote	Unlikely, but possible to occur
Improbable	So unlikely, it can be assumed occurrence may not be experienced

Statistical, Historical Performance Measures

While statistical, historical data is an "after-the-fact" indicator and while the data does not give hazard-specific direction concerning the actions that should be taken for risk reduction, it does provide broad, macro, and meaningful measures of safety performance and serves as a predictor of the future. No matter what criticisms are offered by safety professionals about the shortcomings of historical performance measures, it would be folly for them to ignore the reality that managements usually set goals for improvement that are based on previous results. Historical measures to be discussed here are workers compensation experience rating, workers compensation costs, and OSHA rates.

Workers Compensation Experience Rating

The workers compensation experience modification rating system is based on actuarial science. In an undated publication titled "The ABC's of Experience Rating" (7) issued by the National Council on Compensation Insurance, these statements appear:

> Simply stated, experience rating is a procedure utilizing past insurance experience of the individual policyholder to forecast or predict future losses.
>
> In workers compensation experience rating, the actual characteristics of the individual employer are determined over a period of time, usually three years. This experience is then compared with the average as reflected by the

Table 20.2. Severity of Consequence–
Possible Personal Injury Categories

Catastrophic	Death, permanent disability
Critical	Severe injury or illness
Consequential	Minor injury or illness
Negligible	No injury or illness

manual rate or rates which apply to the employer's business. If the employer has lower than average costs, then a comparable rate credit is awarded, while for a higher than average experience, a debit rate is applied.

What does all this mean? Actuaries have established that workers compensation claims costs, payrolls, and rates for an insured's occupational classes, over time, form a statistical base from which to compute expected claims experience.

Experience rating is mandatory for all employers who buy workers compensation insurance from insurance companies. For those employers, experience rating is one of the historical performance measures that can be used, *cautiously,* as an indicator of the quality of safety in place. Self-insured companies would not have workers compensation experience modifications.

I do not suggest that experience rating is an absolutely accurate indicator of the future. There are many checks and balances in the system, but its preciseness and its sensitivity to change are questioned. And the system is influenced by one-time events that affect the computations for several years. An example would be the shutdown of an operation that resulted in a rash of claims. Nevertheless, *with caution,* an experience rating modification can be used as one indicator of the quality of safety in place, and as a predictor of the future—along with other measures.

For at least one industry, the experience modification system is particularly significant. In the construction industry, it is common practice for a bidder's experience modification to be one of the qualification and acceptance criteria used by a general contractor.

Workers Compensation Costs

If it were possible to obtain *accurate and current* workers compensation claims costs, a trending of that data would be the best performance indicator for most companies, since the data would be expressed in financial terms. That is a language that managements understand. Unfortunately, the actual cost of individual claims is not immediately known. For claims reported in a given year, the actual costs may not be finalized for as long as six or more years.

Still, workers compensation costs present opportunities for computation as performance measures. With the help of insurance personnel, fairly good estimates of the expected claims costs for an ensuing year can be made, providing there are no catastrophic occurrences. With that data, interesting and useful performance indicators can be computed. Some examples follow:

1. If the workers compensation costs are stated as a fixed number, not subject to later revision because of changes in dollar reserves for particular claims, doubt concerning their reliability will be less frequently expressed. To get to a fixed cost number, consider this exercise. Assume that a company's culture has been pretty well set, that things have been somewhat stable, the quality of safety has not significantly changed, and the company's hours worked represent a fairly sound statistical base.

Then the workers compensation cost figure to be used in this exercise is the total of claims paid in one year, *regardless of when the injury or illness was reported.* That figure, once established, would not change. For successive years, those figures will indicate trends, favorably or unfavorably. Allowances are necessary for increases in workers compensation benefits and for inflation when comparing cost data for successive years.

2. In a real-world situation, a safety director obtained numbers that management agreed were reasonable estimates of how annual workers compensation costs were trending. She then determined that her company's OSHA recordable and lost workday case rates were about one fifth that of its industry average. Doing a simple extrapolation, she computed that the company's annual workers compensation costs would be $40,000,000 higher if its OSHA rates were at the industry average. Very impressive.

3. Where it might get attention, I suggest experimenting with a computation that establishes annual workers compensation costs per hour worked, using as a base the total dollars for claims paid in a year, regardless of when the claim was reported. In one company where the costs were computed to be $0.45 cents an hour, the CEO jested that he would kill for a 20 percent reduction. Why such a response? A savings of $0.09 an hour, for 2,000 hours worked, translates to $180 per employee, per year. For 30,000 employees, that becomes $5,400,000. With 40,000,000 shares of stock outstanding, a 20 percent reduction in workers compensation costs improves earnings by $0.135 per share. Very impressive.

But, using workers compensation costs per hour won't get much attention in a company in which the culture has required and achieved exceptionally good safety performance. In a company that has operations the public would consider high hazard, the annual workers compensation cost per employee has recently ranged a bit plus or minus of $100. That computes at $0.05 per hour. A 20 percent reduction nets the employer $0.01 per hour, or $20 per employee per year. Not very impressive.

4. Over time and when the exposure base is sufficiently large, comparisons of workers compensation costs and OSHA statistics, with other companies in the same industry or those considered to have comparable risks, should have a positive and linear relationship. In a benchmarking process, an individual shared data with a company that had an OSHA recordable rate of 0.9, which was about one seventh of his own company's record. He expressed doubt about the validity of the 0.9 OSHA rate, but became a believer when workers compensation cost trendings were compared. There was a match between the OSHA records and the workers compensation costs of the two companies. That should not be an isolated case; such relationships should commonly exist.

5. Another measurement system that could be of value is the rate, recorded over time, of workers compensation claims reported per 200,000 hours worked. Data for such a graph would readily be available. While the rate of workers compensation claims reported will not match precisely a similar graph showing the OSHA reportable rate or the lost workday case rate, great differences should be a subject of concern.

For example: in a very large company, where an edict has been issued that a dramatic reduction is to be achieved in the OSHA recordable rate, the reduction is being accomplished; but the rate of workers compensation claims reported does not match the reduction in the OSHA recordable rate.

Such a graph would be a trend indicator and could provide an alert concerning situations that need attention.

OSHA Rates

Without question, there are inconsistencies, even within companies, in classifying and recording OSHA statistics. Still, if the inconsistencies in the reporting system remain constant, the data produced will serve as useful performance and trend indicators.

The actuarial premises on which the workers compensation experience rating system was developed give credibility to OSHA recordable and lost workday case rates as measures and predictors of safety performance, with these qualifications: the statistical base, the hours worked, on which the records are developed has to be large enough; and low-probability/severe-outcome risks may not be encompassed within the experience base.

In statistical circles, a commonly used term is "the myth of small numbers." Assume that an employer had one hundred employees who worked 200,000 hours in a year. For the employer's industry, the average OSHA recordable rate is eight, and the employer's experience was right on average. For some months in the year, no recordable incidents would have occurred. If the incident distribution is random, more than one incident could have occurred in more than one month. As the year progressed, or at the end of a year, presuming that this employer's OSHA record accurately represented the quality of safety in place would be mythical. Statistically, the exposure sample is not large enough to be credible.

For an entity that small, a combination of statistical measures (the experience modification, a trend chart on which recordings are made of totals of three years of OSHA data, dropping a quarter of a year and adding a new quarter) and qualitative performance measures would be necessary.

Now, do some wild speculation with me, and consider whether what follows begins to have credibility.

An employer has 500 employees who work 1,000,000 hours a year. Before OSHA, ANSI Z16.1—Method of Recording and Measuring Work Injury Experience (8)—was the prevailing recording and measuring guide, and the base for computations was 1,000,000 hours. Computations based on that unit of exposure had some, but not exceptional, reliability.

Do the OSHA statistics, the recordable case rate and the lost workday case rate, for an exposure of 1,000,000 hours have a confidence level of, say 68.27 percent, as measures of the quality of safety performance? An entity of this size would more than likely purchase workers compensation insurance and have an experience modification as an additional measure.

Now move the hours worked to 10,000,000. Will the confidence level of OSHA statistics as performance indicators be as high as 95.44 percent? At 20,000,000 hours, how about a confidence level of 99.73 percent? At 40,000,000 hours, would you go for 99.9937 percent? We could go on with this exercise, increasing the size of the exposure base, and thereby the validity of the OSHA statistics. But, no matter how large the statistical base became, we could never conclude that OSHA rates, nor any other historical data, had a 100 percent confidence level as performance measures or predictive indicators.

Thus, even the largest employers should also be using hazard-specific and qualitative performance measures, for which discussion follows.

Hazard-Specific Identification Measures

A criticism offered of historical, after-the fact data (of outcome statistics) is that such measures are not hazard-specific; that they do not identify incident causal factors without additional analysis. That's so. If safety professionals want to identify hazard-specific situations that may be predictive and give direction to the actions that should be taken to reduce risk, they will have to do a great deal of analysis.

Some of those analyses will be based on historical data, some will be made of specific tasks, and some will seek to identify hazards for which the potentials have not yet been realized.

All such initiatives—such as incident analysis, task analysis, Significant Potential Surveys—produce outcomes that are predictive in one respect. They identify those aspects of technology (the characteristics of things) or the actions or inactions of persons which, if not eliminated or controlled, may be the causal factors for additional hazards-related incidents. They are qualitative measures in that they give an indication of the extent to which policies, standards, and procedures, or their implementation, have been less than adequate, and thereby, how extensively hazards have been integrated into the workplace.

Incident Analysis

Patrick R. Tyson was right in his article "OSHA Proposes Changes to Recordkeeping Rules" (9) when he wrote this:

> After all, a company's injury and illness experience should be a cornerstone of its safety and health program. More simply, it is hard to design a program to keep people from getting hurt if you don't know what's hurting them.

To "know what's hurting them" requires effective incident analysis. Identifying "what's hurting them" provides predictive information on what may hurt employees in the future. To "know what's hurting them," a safety professional must understand incident causation and craft an analysis system that effectively identifies casual factors. A balance must be built into such a system that gives the necessary emphasis to hazards that derive from design and engineering, operations management, and task performance practices.

If incident investigations and analyses are done poorly, the conclusions drawn will result in misdirected actions to reduce risk. Also, the conclusions will not be accurate predictors of the future. Incident analysis is a subject that should be taken seriously and considered as a predictive exercise—treated as though it is one of the cornerstones of a safety and health initiative.

Help in crafting an incident investigation and analysis system can be found in Chapter 6 ("Observations on Causation Models for Hazards-Related Incidents"), Chapter 7 ("A Systemic Causation Model for Hazards-Related Occupational Incidents"), Chapter 8 ("Incident Investigation: Studies of Quality"), and Chapter 9 ("Designer Incident Investigation").

Task Analysis

While incident analysis is after-the-fact, task analyses can be made after-the-fact or before-the-fact. Task analyses may be called job hazard analysis, job safety analysis, or total job analysis. Whatever the name given to the process, the results of task analyses are qualitative, and can be predictive. This extract is from the *Handbook of Occupational Safety and Health* (10) (edited by Lawrence Slote):

> Job Safety Analysis (JSA) or Job Hazard Analysis (JHA), a systematic study of work procedures, is utilized by those firms who desire to identify and control hazards before such hazards result in injury.

In *Management Guide to Loss Control* (11), by Frank E. Bird, Jr., a chapter on "Proper Job Analysis and Procedures" extends the purposes of a task analysis to include aspects other than safety. Bird says that:

> The proper job analysis and standard job procedure chapter is based on the concept that all elements of a worker's job, such as quality, production, safety and health, are inseparable. Any one or all can affect the others, and to consider them as separate elements when teaching a worker to do his job is

to invite the confusion and misunderstanding that leads to downgrading incidents.

I suggest that safety professionals give task analysis a greater emphasis as a measure of the quality of safety in place for a task, and as a qualitative predictor of the probability of hazards-related incidents occurring. Task analyses are to define hazardous or inefficient work procedures.

If a safety professional really wants to know what hazards may create problems tomorrow, task analysis is a highly effective way to identify them. Of course, the process would culminate in proposals for the appropriate preventive actions. The ancillary benefits are considerable, since many people can be trained through the task analysis process to identify hazards and learn how to seek their elimination or control.

Surveys to Identify Potential Causal Factors

The survey system to be discussed here had its origins in what has been known as The Critical Incident Technique. Although the technique has not been broadly used, it has real possibilities in identifying hazards before their potentials are realized. Also, application of such a system would build a body of predictive, hazard-specific knowledge.

In *The Measurement of Safety Performance* (3), William E. Tarrants devotes a chapter to "The Critical Incident Technique as a Method of Identifying Potential Accident Causes." His comments on the origin of the technique and its purposes follow:

> . . . the critical incident technique . . . is regarded as an outgrowth of studies in the Aviation Psychology Program of the U.S. Army and Air Force in World War II. It is an accident study method in which an interviewer questions a number of persons who have performed particular jobs and asks them to recall within a specified time period unsafe acts and/or conditions they have committed or observed.
>
> The persons are selected on a stratified random sampling basis, with stratifications designated according to the type of exposure, quantity of exposure, degree of hazard present, and other criteria considered important to the representativeness of the sample. The objective is to discover causal factors that are critical, that is, that have contributed to an accident or potential accident situation. The unsafe acts and unsafe conditions identified by this method then serve as the basis for the identification of accident potential problem areas and the ultimate development of countermeasures designed to control accidents at the no-loss stage.

Willie Hammer wrote this in his *Handbook of System and Product Safety* (12).

> The Critical Incident Technique is one means by which previously experienced difficulties can be determined by interviewing persons involved. It is based on collecting information on hazards, near misses, and unsafe conditions and practices from operationally experienced personnel.
>
> The technique consists of interviewing personnel regarding involvement in accidents or near accidents; difficulties, errors, and mistakes in operations; and conditions that could cause mishaps.

A modification of The Critical Incident Technique was published by the Division of Safety, Standards, and Compliance of the United States Energy Research and Development Administration in 1976. Because the method was also to be applied in the nuclear industry, where the terms "critical" and "incident" have their own connotations, a new name was used—"Reported Significant Observation (RSO) Studies" (13). In a paper with that title, it was said that "RSO was formally recognized as a significant hazard reduction tool."

I propose a different name for the technique. While I suggest the title "Potential Causal Factors Surveys," any name chosen by those designing such a system would be as suitable. But why make so much of the idea on which the Critical Incident Technique is based? A system that seeks to identify potential causal factors before their potentials are realized would serve well in attempting to avoid low-probability/severe-consequence events. Earlier I wrote that:

- the causal factors for such events are usually complex and may be different from the causal factors for events that occur frequently;
- low-probability incidents often involve unusual or nonroutine work, high energy sources, nonproduction activities, and certain construction situations.

Therefore, safety professionals must have within their endeavors a separate and distinct activity to seek those hazards that present the most severe injury or damage potential, so that they can be given the priority consideration they require. Making Potential Causal Factors surveys would serve that purpose.

In *Profitable Risk Control* (14), William W. Allison makes a good case when he suggests that we learn how to identify and concentrate limited time and money on controlling the significant, few high-potential risks before they become costly risks. This is how Allison defined a high-potential risk:

> A high potential risk is any situation, practice, procedure, policy, process, error, occurrence or accident which may or may not have resulted in loss or harm but can result in significant harm to people, product, services, equipment, facilities, or property.

A system designed to make Potential Causal Factors Surveys, a qualitative measurement system, would produce data that could be predictive of the future.

Scoring Individual Safety Processes and Safety Audits

Some have suggested that predictive measures could be developed by scoring the accomplishments of individual safety activities, such as training, inspections, compliance with standards, or exposure assessment surveys.

If an activity truly related to an entity's risks, a scoring system could be developed that had predictive values. But, measuring systems that produce numerical indicators such as the number of persons trained, the number of inspections made, or the number of health exposure surveys completed would be of little value if those processes did not relate to the hazards in place. As an example, I cite this personal experience about training.

For many years, I led training programs on how to lift safely. I now know that in most of those situations, training did not address the problem. The design of the work was overly stressful for a large percentage of the working population, and that was what needed attention. If a performance measure was to give a score based on the number of training sessions I conducted or the number of people trained, it would give a false indication of effectiveness.

There are a variety of numerical scoring systems for safety audits. But, for some safety audit subjects it is not possible to imagine suitable subjectively concluded scores. Suppose the following management practice had been specified in a company's issued procedures, and a safety professional tried to put a quality scoring on it while making a safety audit.

> Designs for new facilities and modifications of existing facilities, and the tasks to be accomplished, are to be reviewed to assure that hazards are identified and analyzed, and avoided, eliminated, or controlled so that risks are at a practical minimum.

Could a scoring system produce meaningful statistics that would rate such a management practice and be predictive of the future? I don't think so.

Nevertheless, safety audits, properly conducted, provide highly effective qualitative performance measures (see Chapter 17).

Conclusion: Being Realistic

Because of the impossibility of knowing all risks on an anticipatory basis, it is folly to expect that a perfect performance measurement system can be developed. Also, because of the limitations of humans, it is not possible to attain a risk-free environment and assure zero occurrences of hazards-related incidents.

Several years ago Motorola established a quality performance level known as Six Sigma. In its literature on the program, this appears:

> As we have seen, even the most well-controlled processes experience shifts of the process mean due to changes in equipment, operators, environmental conditions, and incoming materials. Such shifts can be as much as 1.5 standard deviations.
>
> Because process shifts do normally happen, a Six Sigma quality level in Motorola's manufacturing operations has been defined as 99.99966 percent. In other words, 99.99966 percent of all parts produced are defect-free. This is the same as saying that only .00034 percent of parts produced—3.4 parts of every million—are defective.
>
> This is just about as close to perfection as human beings can get.

Since we are not capable of knowing, avoiding, or controlling all risks on an anticipatory basis, should we realistically concede, as the Motorola statement implies, that developing perfect predictive measuring systems and achieving zero hazards-related incidents is unattainable?

Assume that an OSHA recordable incident is the defect to be measured. How does 3.4 defective parts per million relate to an OSHA recordable incident rate? OSHA rates are computed from a base of 200,000 hours worked. To be at an OSHA recordable rate of 3.4 incidents per million hours, the computed rate using a 200,000 hour base would 0.68. Is that attainable in operations considered moderate or high hazard? I could identify only one such company, with a great many hours worked, that published an OSHA recordable rate slightly lower than 0.68 for 1995: the next lowest record located for what some would consider a high-hazard operation was slightly higher than twice 0.68.

For moderate or high hazard entities, are those achievements "about as close to perfection as human beings can get"?

I find the following comments about the practical limitations on knowing all there is to know about risks, before or after the fact, to be illuminating. They are taken from "The Crash Detectives" by Jonathan Harr (15), an article on airplane crashes and the work of investigators to identify their causal factors. It appeared in the August 5, 1996, issue of *The New Yorker*:

> Modern jet airplanes are designed with highly redundant systems, which make accidents highly improbable. When they do occur, they are usually the result of a concatenation of discrete events—of mechanical or human failures—any one of which by itself would not likely cause a catastrophe. It is this unforeseen sequence of events, resulting from what accident theorists call the "tight coupling" of complex interacting systems, that causes accidents.
>
> The F.A.A. announced a goal of zero accidents early last year. This, of course, sounds like a laudable goal. Perhaps it is just meant to reassure a

> worried public, for almost no one—least of all the investigators at the Major Investigations Division of the [National Transportation] Safety Board—believes that such a goal is attainable.

Although I quote from Motorola, I do not attest that attaining a record of 3.4 defects per million parts produced is the best that humans can do. However, I do say that it is not possible for humans to design perfect performance measurement systems that detect 100 percent of the risks, before or after incidents occur.

REFERENCES

1. Manuele, Fred A. "How Do You Know Your Hazard Control Program Is Effective?" *Professional Safety,* (June 1981).
2. Bailey, Charles W., and Dan Petersen. "Using Perception Surveys to Assess Safety System Effectiveness," *Professional Safety.* (February 1989).
3. Tarrants, William E. *The Measurement of Safety Performance.* Garland Publishing: New York, 1980.
4. Hanayas, Shigao. "One Method For Evaluating Safety Performance in Working Places." In Proceedings of the 8th Congress of the International Ergonomics Association, [Tokyo] August 1982.
5. Bird, Jr., Frank E., and Robert G. Loftus. *Loss Control Management.* Loganville, GA: Institute Press, 1976.
6. *Program Performance Measures.* Chemical Manufacturers Association, Washington, D.C., 1995.
7. *The ABC's of Experience Rating.* National Council on Compensation Insurance, New York.
8. Z16.1—*Method of Recording and Measuring Work Injury Experience.* American Standards Association, New York, 1973.
9. Tyson, Patrick R. "OSHA Proposes Changes to Recording Rules." "*Safety+Health.*" (May 1996).
10. Slote, Lawrence, ed. *Handbook of Occupational Safety and Health.* John Wiley & Sons, Inc., New York, 1987.
11. Bird, Jr., Frank E. *Management Guide To Loss Control.* Loganville, GA: Institute Press, 1978.
12. Hammer, Willie. *Handbook of System and Product Safety.* Englewood Cliffs, NJ: Prentice-Hall, 1972.
13. *Reported Significant Observation (RSO) Studies.* Division of Safety, Standards, and Compliance of the United States Energy and Development Administration. National Technical Information Service, Springfield, VA, 1976.
14. Allison, William W. *Profitable Risk Control.* Des Plaines, IL: American Society of Safety Engineers, 1986.
15. Harr, Jonathon. "The Crash Detectives." *The New Yorker,* August 5, 1996.

Index

Academic and skill requirements, 35–44
  curriculum standards and BCSP safety fundamentals examination, 39–40
  curriculum standards matching educational needs, 36–37
  reviewing curriculum standards, 37–39
Acceptable risk, 182–183
Accountability, establishing, 238–239
Analyses
  incident, 260–261
  parameters, 196
  qualitative hazard, 185
  task, 261–262
ASHRAE (American Society of Heating, Refrigeration and Air-Conditioning Engineers), 167
ASSE (American Society of Safety Engineers), 35–36, 48–49, 52
Auditing system, safety, 231–232
Audits
  purpose of, 220
  safety, 25–26
  safety, health, and environmental, 218–225
    assessing senior management commitment, 220–221
    audit effectiveness of, 223–224
    definition of audits, 219–220
    design implications, 221–223
    engineering implications, 221–223
    giving alternate solutions, 223
    prioritizing recommendations, 223
    self-review outline, 218–219
  scoring safety, 264
  self, 99–101

Baldrige, Malcolm, 203–205
BCSP (Board of Certified Safety Professionals), 35, 52
  baccalaureate curriculum standards, 36
  safety fundamentals examination, 39–40
Behavior modification, 19–20
Blame, avoiding placing, 106
Business
  practices, 2–4
  problem of perception, 7–8

Career base, solid, 10–11
Causal defined, 77
Causal factors
  codes, 89–90
  defined, 77
  derivation of, 71
  designing, 67–68
  development, 82
  multiple, 107

reference for, 107–113
surveys to identify, 262–263
Causation
defined, 77
incident, 23–24
models, 56–73
all incident, 57–58
based on Heinrichian principles, 61–63
causal factors, derivation of, 71
causal factors development, 82
choosing, 99
defining hazards-related incidents, 68–71
defining incident phenomenon, 60–61
definitions, 58–59, 76–77
design decisions, 81
design management, 79–80
designing causal factors, 67–68
diversity of thinking, 58
error-provocative work, 65–66
hazard analysis, 79
for hazards-related incidents, 56–73
hazards-related incidents, 82
hazards-related occupational incidents, 74–82
inappropriate terms, 63–64
need for accepted, 59–60
observations on, 56–73
operations management, 80–81
operations practices, 81
organizational culture defined, 78–79
premises on which based, 77–78
professional safety practice, 56–57
recommended readings, 68
relationships, 81
risk assessment, 79
supporting discussion, 78
supporting premises, 61
systemic, 74–82
systems, procedures, and work environment, 64–65
task performance, 81
unsafe acts, 63–64
unsafe conditions, 63–64
Certification requirements, maintaining rigid, 52
Change, management of, 241–242
Codes
causal factor, 89–90
incident type, 90
injury type, 90
Commitment, management, 238
Committees, safety, 243–244
Communication systems, 240
Conduct, adhering to accepted standard of, 53
Consequences, assessing severity of, 196
Contract specification check list, 165–169
chemical, biological, radiological, 166
electrical safety, 165
emergency safety systems, 166
environmental, 168
ergonomics workstation and work methods design, 167–168
fire protection, 165–166
mechanical safety, 165
pressure vessels, 167
ventilation, 167
walking/working surfaces, 165
wastewater, 169
Controls
environmental, 246–247
statistical process, 216
thought processes for, 162–164
Costs
remediation, 197
workers compensation, 257–259
Crosby, Philip B., 209–211

CSP (Certified Safety Professional), 52
Culture defined, 235–238
Cultures, organizational, 18, 78–79, 197
Curriculum standards
  and BCSP safety fundamentals examination, 39–40
  matching educational needs, 36–37
  reviewing, 37–39

Decisions
  design, 81
  matrix, 192–193
Defects, zero, 209–211
Deming, Edwards, 206–208, 214–216
Department managers, 172–173
Design
  decisions, 81
  engineers, 173
  management, 79–80
Devices
  incorporating safety, 161
  providing warning, 161
Disaster planning, 249
Domino sequence, 61–62
Drucker, Peter, 9–10

Education, curriculum standards, 36–39
EHS (Environmental Health and Safety), 5
Electrical safety, 165
Elimination, thought process for, 162–164
Emergency planning, 249
Emergency safety systems, 166
Employee involvement, 127–128, 228, 246
Energy, expenditures of, 177–178
Engineering, knowledge and skill requirements and, 41–42
Engineering, human factors, 242
Engineers, design, 173
Environmental affairs, 13
Environmental controls, 246–247
Environmental professionals, 173
Ergonomics
  applied, 11–12, 114–133
  defining, 115–116
  including productivity and cost effectiveness, 116–118
  manual material handling, 118–120
  participating in design process, 120
  professional recognition, 118
  program management recommendations, 126–132
    employee involvement, 127–128
    management commitment, 127–128
    program elements, 129–132
  recommended reading, 121
  significance of, 21–22, 114–115
  standard and OSHA, 124–126
  undertaking initiative, 122–124
  workplace safety, 114–115
  workstation and work methods design, 167–168
Error, on human, 152–154
Error-provocative work, 65–66
Experience rating, workers compensation, 256–257
Exposures
  defining, 177
  describing, 196

Facilities
  first aid, 249
  medical, 249
Fads, management, 4–7
Failure modes, 196
Fire protection, 165–166
First aid facilities, 249

Goals
  participants in achieving management, 8–9
  safety professionals achieving, 7–8

Government regulations, compliance with, 249
Gryna, Frank M., 206–208, 214–216

Haddon, William, 178–179
Haddon's unwanted energy release and hazardous environment concept, 178–179
Hazard
  determining probability of, 196–197
  prevention and control, 229
  probability, 193–194
Hazard analysis, 79
  is most important safety process, 83–84
  making, 192
  OSHA requirements, 198–199
  what if, 172
Hazard analysis and risk assessment, 188–201
  achieving acceptable risk levels, 194–195
  appropriation requests, 164
  conducting, 196–198
    assessing severity of consequences, 196
    considering failure modes, 196
    defining risk, 197
    describing exposures, 196
    determining probability of hazard, 196–197
    developing remediation proposals, 197
    establishing analysis parameters, 196
    identifying hazards, 196
    influence of organizational culture, 197
    initiating an assessment system, 198
    priority setting, 197
    remediation costs, 197
    risk ranking, 197
  conducting a hazard analysis/risk assessment, 196–198
  contract specifications, 164–169
  developing decision matrix, 192–193
  emphasizing, 188–189
  hazard probability, 193–194
  hazard severity, 193
  OSHA's hazard analysis requirements, 198–199
  project reviews, 164
  resources on, 199–200
  severity potential priority, 189–191
  subjectivity in, 191
Hazard avoidance, thought process for, 162–164
Hazardous conditions, 111
Hazards, 15–16
  acquiring additional knowledge of, 177
  control of occupational health, 246
  defined, 30, 76
  defining, 176
  identification, 176–177
    defining exposures, 177
    defining perils, 177
  identifying, 196
  management, 33
  relating to risks, 176
  review committee, 172
  and risks
    comments on, 175–187
    expenditures of energy, 177–178
    Haddon's unwanted energy release and hazardous environment concept, 178–179
    qualitative hazard analyses and risk assessments, 185
    on significance of MORT, 180–181
    understanding of, 254–256

severity of, 193
significance of, 176
Hazards-related incidents; *See* HAZRIN
HAZRIN (hazards-related incidents), 56–73, 82
creation of, 61
and decision-making, 65
defining, 68–71, 76
Health
hazards, control of occupational, 246
professionals, 173
responsibility for, 13
Heinrich, H.W., 61–63
Heinrichian principles, 61–63
Historical performance measures, 256–260
Human error, 152–154
Human factors engineering, 242

Incident analysis, 260–261
Incident causation, 23–24
Incident data, use of, 248
Incident investigations, 83–92
adopting from what has been learned, 95–96
being sympathetic toward supervisors, 95
choosing a causation model, 99
data gathered for study, 84–86
comments by safety professionals, 86–90
highlights, 84–85
methodology and scoring, 85–86
defining terms, 102
designer, 93–113
designer incident investigation reports, 107–113
corrective actions, 107–113, 110–112
corrective actions to be considered, 112
hazardous conditions, 111
job procedure particulars, 110–111
management and supervisory aspects, 111
personal protective equipment (PPE), 111
reference for causal factors, 107–113, 110–112
selection of subjects, 108–110
work methods considerations, 110
workplace design considerations, 110
establishing purposes of, 101–102
fact finding, 106
hazard analysis, 83–84
immediate actions to be taken, 105
informing on incidents to be investigated, 102–103
investigation, 103–104
making a self-audit, 99–101
multiple causal factors, 107
notification procedures, 105–106
quality of, 99–101
reality observations, 94–95
reporting responsibility, 103–104
resources, 97–99
studies of quality, 83–92
teams, 104–105
techniques, 106–107
training in, 112–113
Incident phenomenon, defining, 60–61
Incidents
defined, 76
hazards-related, 56–73, 82
major, 103–104
outcome of, 82
reporting, 247–248
type codes, 90
Information systems
communication and, 240
technical, 240
Injury type codes, 90

Inspection programs, 247
Investigations, incident, 93–113

Juran, Joseph M., 206–208, 214–216

Knowledge and skill requirements, 40–43
  applied sciences, 42–43
  engineering, 41–42
  legal/regulatory aspects, 43
  management, 42
  professional conduct and affairs, 43
Knowledge society, 9–10

Leadership, 19–20
Logic, and risk management decisions, 184–185

Maintenance
  designing for safety during, 154–155
  preventive, 243
Major incident, 103–104
Malcolm Baldrige National Quality Award, 203–205
Malcolm Baldrige National Quality Award criteria
  business results, 204
  consumer focus and satisfaction, 204
  corporate citizenship, 205
  design and introduction of products and services, 205
  high performance work systems, 205
  human resource development and management, 204
  human resources results, 205
  information and analysis, 204
  leadership, 204
  leadership system and organization, 205
  process management, 204
  public responsibility, 205
  senior executive leadership, 205
  strategic quality planning, 204
Management, 42
  of change, 241–242
  commitment, 78–79, 127–128, 228, 238
  design, 79–80
  fads, 4–7
    history of, 5–7
    theory X assumptions, 5
    theory Y assumptions, 6
  hazards, 33
  operations, 80–81
  quality, 12–13, 202–217
  and risk understanding, 182
Managers
  department, 172–173
  project, 172
Material handling, manual, 118–120
Matrix, developing decision, 192–193
Measurement, performance, 248, 251–266
Mechanical safety, 165
Medical facilities, 249
Model defined, 77
Modes, failure, 196
MORT (management oversight and risk tree)
  and assumption of occurrence of incidence, 98
  bases for, 70, 74, 177
  definition of an accident, 64
  energy releases, 68
  framework, 69
  and hazard analysis, 83–84
  and hazard identification, 79
  significance of, 180–181
  and supporting discussion, 78
  and system safety
    accomplishments, 131
    technology, 97
Motorola, Six Sigma Program at, 211–212

NFC (National Fire Codes), 165
Notification procedures, 105–106

Observations, reality, 94–95
Occupational health hazards, control of, 246
Operating procedures and training, developing, 161–162
Operations
  management, 80–81
  practices, 81
Organizational culture, 18, 78–79, 197
OSHA (Occupational Safety and Health Administration)
  accident models, 57–58
  and ergonomics standard, 124–126
  hazard analysis requirements, 198–199
  history, 226–227
  industry safety and health program standard, 225–234
  investigation methodologies, 57–58
  meeting requirements for hazards analysis, 136
  proposed safety and health program standard, 230
  rates, 90–91, 259–260
  reviews of incident investigations and, 122
  safety and health program guidelines, 227–230, 232–234
    employee involvement, 228
    hazard prevention and control, 229
    management commitment, 228
    safety and health training, 229–230
    scope and application, 227
    worksite analysis, 229
  safety management audit system, 225–234
  work on a safety auditing system, 231–232

Parameters, analysis, 196
Perception, 7–8
Performance measurement, 248, 251–266
  being realistic, 264–266
  hazard-specific identification measures, 260
  ideal safety performance measurement, 253–254
  incident analysis, 260–261
  literature review, 252–253
  scoring individual safety processes, 264
  scoring safety audits, 264
  significance of performance measures, 253
  statistical, historical performance measures, 256–260
  surveys to identify causal factors, 262–263
  task analysis, 261–262
  understanding of hazards and risks, 254–256
Performance measures, 24–25
  OSHA rates, 259–260
  statistical, historical, 256–260
  workers compensation
    costs, 257–259
    experience rating, 256–257
Performance, task, 81
Perils, defining, 177
Planning
  disaster, 249
  emergency, 249
Policy/procedure statement, model, 169–170
PPE (personal protective equipment), 111, 131
Practices
  business, 2–4
  operations, 81
Pressure vessels, 167
Preventive maintenance, 243

Priority setting, 197
Probability, hazard, 193–194
Procedure guide, model, 170–173
  administrative procedures, 173
  definitions, 171
  hazard review committee, 172
  installation review, 171–172
  pre-capital review, 171
  purpose, 170
  responsibilities, 172
    department managers, 172–173
    design engineers, 173
    environmental professionals, 173
    health professionals, 173
    project managers, 172
    safety professionals, 173
  scope, 171
  what if hazard analysis, 172
Procedures, notification, 105–106
Process controls, statistical, 216
Profession, on becoming a, 45–55
  action subjects, 54
  literature review, 45–47
  requirements, 47
    achieving recognition as a profession, 51
    adhering to accepted standard of conduct, 53
    developing a common language, 50–51
    establishing a theoretical and practical base, 47–50
    having a professional society, 53
    maintaining rigid certification requirements, 52
    obtaining societal sanction for professionalization, 53
    promoting and supporting research, 52
Profession, hallmarks of a, 47
Professional safety practice, 56–57
Professional society, having a, 53
Professionalization, obtaining societal sanction for, 53
Professionals
  environmental, 173
  health, 173
  safety, 173
Programs, inspection, 247
Project managers, 172
Proposals, remediation, 197
Protection, fire, 165–166
Purchasing standards, 242–243

Qualitative hazard analyses, 185
Quality
  management, 12–13, 202–217
    emphasizing system and process design, 214–216
    learning from Deming, Juran, and Gryna, 206–208
    Malcolm Baldrige National Quality Award, 203–205
    Six Sigma Program at Motorola, 211–212
    on statistical process controls, 216
    theoretical ideal, 203
    why quality and safety initiatives fail, 212–214
    zero defects, 209–211
  studies of, 83–92
Quantitative risk analyses, 184–185

Rates, OSHA, 259–260
Ratings, workers compensation experience, 256–257
Reality observations, 94–95
Regulations, government, 249
Relationships, 81
Remediation costs, 197
Remediation proposals, 197
Reporting
  incident, 247–248
  responsibility, 103–104
Residual risks, 195

Resources, utilizing, 22
Responsibilities
  investigation, 103–104
  reporting, 103–104
Risk analyses, quantitative, 184–185
Risks
  acceptable, 182–183
  assessments, 79, 185, 188–201
  defined, 29–30, 181–182, 197
  design for minimum, 161
  estimates that result in money numbers, 186
  impact of, 194
  levels, achieving acceptable, 194–195
  management decisions and logic, 184–185
  ranking, 197
  reduction advice, 183
  residual, 195
  role of safety, 181–182
  and safety, defining, 16
  understanding management's view of, 182
  validity of quantitative risk analyses, 184–185

Safe practice standards, 247
Safety, 13
  audits, 25–26, 231–232, 264
  committees, 243–244
  defining practice of, 16–17, 27–34, 31–33
    baffling and nondescriptive titles, 30–31
    definition, 32–33
    hazards management, 33
    knowledge and skill requirements, 33–34
    major elements, 33
    position description, 33
  devices, 161
  electrical, 165
  health, and environmental audits, 218–225
  and health training, 229–230
  management
    of change, 241–242
    commitment, 78–79, 238
    communication systems, 240
    compliance with government regulations, 249
    control of occupational health hazards, 246
    culture defined, 235–238
    design and engineering, 240–241
    direction, 238
    disaster planning, 249
    emergency planning, 249
    employee involvement, 246
    environmental controls, 246–247
    ergonomics, 242
    establishing accountability, 238–239
    first aid facilities, 249
    hazard analysis, 241
    human factors engineering, 242
    incident reporting, 247–248
    information systems, 240
    inspection programs, 247
    investigation, 247–248
    involvement, 238
    medical facilities, 249
    performance measurement, 248
    preventive maintenance, 243
    purchasing standards, 242–243
    recording, analysis, and use of incident data, 248
    reflection of an organization's culture, 235–250
    risk assessment, 241
    safe practice standards, 247
    safety committees, 243–244
    safety organization and staff, 239–240
    successful, 235–250

supervisory participation and accountability, 244
technical information systems, 240
training, 244–246
mechanical, 165
organization and staff, 239–240
performance, 251–266
practice, 56–57
principles for practice of, 15–26
behavior modification, 19–20
defining the practice of safety, 16–17
defining risk and safety, 16
design and engineering, 20–21
hazards, 15–16
incident causation, 23–24
leadership, 19–20
organizational culture, 18
performance measures, 24–25
safety audits, 25–26
setting priorities, 22
significance of ergonomics, 21–22
system safety, 21
theoretical ideal for safety, 17–18
training, 19–20
utilizing resources effectively, 22
processes
hazard analysis is most important, 83–84
scoring individual, 264
role of, 181–182
systems, 21, 134–144, 166
theoretical ideal for, 17–18
transitions affecting practice of, 1–14
and business practices, 2–4
major influences, 2
management fads, 4–7
workplace, 114–115
Safety, designing for
avoiding introduction of hazards, 162
design requirements, 162–164
guidelines, 157–174
hazard analysis, 164–169
limiting amount of energy or hazardous material, 162–163
model policy/procedure statement, 169–170
model procedure guide, 170–173
modifying shock concentrating surfaces, 164
preventing unwanted energy, 163
preventing unwanted hazardous material buildup, 163
principal resources, 158
protecting people, property, or environment, 164
purpose, 159–162
definitions, 159–160
design for minimum risk, 161
developing operating procedures and training, 161–162
general principles, 159–160
incorporating safety devices, 161
objectives, 160–161
order of design precedence, 161–162
providing warning devices, 161
risk assessment, 164–169
separating release of energy, 163
separating release of hazardous materials, 163
slowing down release of energy, 163
slowing down release of hazardous material release, 163
using less hazardous substitutes, 163
Safety professionals, 145–156, 173
being anticipatory and proactive, 151–152
benefits besides that of safety, 150–151
design implications
a macro view of, 149–150
in the operational mode, 147–148
in the post-incident mode, 148–149
in the pre-operational stage, 147

and design processes, 145–156
designing for safety, 152–155
future of, 145–147
history, 145–147
on human error, 152–154
Sciences, applied, 42–43
Self-audit, 99–101
Severity, hazard, 193
Six Sigma Program (Motorola), 211–212
Skill requirements, 35–44
Societal transition, 9–10
Society, knowledge, 9–10
Standards
purchasing, 242–243
safe practice, 247
Statistical performance measures, 256–260
Statistical process controls, 216
STEP (Sequentially Timed Events Plotting), 97
Study, data gathered for, 84–86
comments by safety professionals
causal factor codes, 89–90
causal factor determination quality, 88–89
cause categories, 89
general information, 87
incident descriptions, 87–88
incident type codes, 90
injury type codes, 90
investigations by teams were superior, 90
report titles, 86–90, 87
observations from a second study, 90–92
Supervisors, being sympathetic towards, 95
Supervisory participation and accountability, 244
Surfaces, walking/working, 165
System safety, 21, 134–144
application outline, 138–139
concepts not widely adopted, 140–141
defining, 136–138
evolution of, 139–140
idea, 137
influences encouraging adoption of, 135–136
promoting use of, 141–142
recommended reading, 142–143
Systems
communication, 240
emergency safety, 166
information, 240

Task analysis, 261–262
Task performance, 81
Teams, incident investigation, 104–105
Technical information systems, 240
Titles, baffling and nondescriptive, 30–31
TQM (total quality management), 212
Training, 19–20, 244–246
Transitions
affecting safety practice, 1–14
societal, 9–10

Validity of quantitative risk analyses, 184–185
Ventilation, 167
Vessels, pressure, 167

Walking/working surfaces, 165
Warning devices, providing, 161
Wastewater, 169
Work, error-provocative, 65–66
Workers compensation
costs, 257–259
experience rating, 256–257
Workplace safety, 114–115
Worksite analysis, 229
Workstation and work methods design, 167–168

Zero defects, 209–211